Bruna Grossl

Analysing the proposal to build bus corridors

Bruna Grossl

Analysing the proposal to build bus corridors

Case study carried out in Joinville-SC

ScienciaScripts

Cover image: www.ingimage.com

This book is a translation from the original published under ISBN 978-613-9-63612-9.

Publisher:
Sciencia Scripts
is a trademark of
Dodo Books Indian Ocean Ltd. and OmniScriptum S.R.L publishing group

120 High Road, East Finchley, London, N2 9ED, United Kingdom
Str. Armeneasca 28/1, office 1, Chisinau MD-2012, Republic of Moldova, Europe
Printed at: see last page
ISBN: 978-620-7-67487-9

ACKNOWLEDGEMENTS

To God for having enlightened and blessed the way for me to get here, for the answered prayers and for the company when I felt alone.

To my fiancé Allan, who was there for me not only in the many happy moments, but also in the moments when I thought about giving up, supporting me and believing in me.

To my family who have always encouraged me to do my best, who have helped me to follow my dreams even though they have kept us apart. Especially my younger sister Carla, who studied with me on several weekends and was patient on the days when I was in a bad mood.

To my supervisor Elisete, who always believed in me and in this work, who had the right words at uncertain times and who became a friend throughout these years of university.

To the teachers on my course who guided and encouraged me throughout university, dedicating their time to preparing me for the new phase that lies ahead.

To my co-workers at SEPUD for learning and for the moments of relaxation, especially Taline and Gilson for helping to carry out this work.

To my friends who have walked alongside me on this journey, who have been willing to listen to my complaints and who have helped me over the years, especially Maria, Larissa Bento, Monique and Germano.

To all those who have contributed to this achievement throughout my life.

My sincere thanks.

SUMMARY

The need to solve urban mobility problems arising from uncontrolled and often unplanned population growth has led to the creation of numerous measures to increase the quality and performance of urban public transport services. One of these is the implementation of exclusive bus lanes, particularly the **Bus Rapid Transit** (BRT) system, which is becoming increasingly popular due to success stories in populous cities, where it has proved to be an attractive and effective option. The aim of this study was therefore to analyse the proposal for the construction of bus corridors made by Joinville City Council and submitted to PAC 2 Mobilidade Médias Cidades, comparing it to the documents Quality Standard BRT 2016 and BRT Manual 2008. As for the methodology, this is an applied study, classified as exploratory and descriptive, in which the technical procedures are characterised by bibliographical and documentary research and a case study. The analyses carried out throughout the work, both in the light of the guiding documents BRT Quality Standard and BRT Manual, and by consulting professionals in the field, indicate that the proposal to build bus corridors has the potential to be efficient and bring improvements to Joinville's Integrated Transport System (ITS). However, they also highlight the lack of studies such as the design of lines and networks, the identification of bottlenecks and traffic light control. The lack of these, coupled with the construction of preferential corridors linking the terminals, will probably jeopardise the quality of the operation of the exclusive bus corridors proposed by Joinville City Hall.

Keywords: Urban public transport. Bus corridors. Operational studies.
BRT system.

SUMMARY

CHAPTER 1

INTRODUCTION

The Brazilian economic crisis has affected urban public bus transport, damaging its performance and consequently its financial health. Between 2014 and 2016, the national scenario saw an 18% drop in the number of passengers transported by urban buses, with direct repercussions on the passenger per kilometre index (IPK), which shows a reduction in the sector's productivity of more than 11%. This reduction generates an economic and financial imbalance due to the fall in revenue and increase in the cost per passenger transported, which has been observed for some years in the main Brazilian cities (NTU, 2017).

The popular demonstrations that took place in several Brazilian cities in 2013 brought to light the problems experienced by urban transport users, highlighting their dissatisfaction with the price of urban public transport fares and the precariousness of the provision of this service (SILVEIRA; COCCO, 2013). The economic and financial imbalance not only compromises the quality of public transport services and the financial management capacity of the operating companies, but can also jeopardise access to these services for the low-income population. This situation is exacerbated during periods of crisis, when family incomes fall (CARVALHO; PEREIRA, 2011).

In the last 60 years, Brazil has experienced a rapid process of urbanisation, according to Carvalho and Pereira (2011), which has contributed to the consolidation of major socio-economic inequalities in the occupation of urban space, where the lower social classes live in more segregated peripheral areas and public transport takes on a central role. However, public transport policies and other economic policies have not been able to contain the growth in public transport costs.

Branco (2013) states that the process of urbanisation highlights the problems caused by the unrestricted use of cars: energy waste, social, environmental and economic burdens, and road saturation. Therefore, since urban mobility is an imperative urban need, it is necessary to rethink the transport model of cities in order to make them more sustainable.

The need to solve urban mobility problems caused by rapid and often unplanned population growth has stimulated the creation of numerous measures to increase the quality and performance of public transport services. Of particular note is the introduction of exclusive lanes for urban public transport, a measure that is often mistakenly referred to as a BRT system. BRT systems have had their potential recognised by successful results in populous cities, proving to be an attractive and effective option. For this reason, there has recently been a growing popularity of these systems, including various experiences in Latin America, justifying the interest in developing this work.

According to Branco (2013, pg. 1),

> [...] implementing an effective public transport system that is compatible with the current economic reality and that provides a high-quality, safe, comfortable and fast service is an ongoing challenge for the bodies responsible. However, it is essential to guarantee an efficient service since, for a considerable percentage of the population, public transport is the only alternative for meeting their mobility needs.

Uncontrolled growth in medium-sized cities has led to demands for improvements in public transport, especially when we look at the drop in public transport users and the migration to individual transport. In recent years, users of the public transport system in Joinville, classified as a medium-sized city, have been dissatisfied with the price of fares and the service provided. The process of urbanisation, in many cases without planning, has led to a poor division of land use, increasing the number of journeys needed for citizens to get around and carry out their daily activities. Encouraging the use of private transport and the poor quality of public transport, as evidenced by recent satisfaction surveys (EMBARQ BRASIL, 2014), lead to congestion, which demonstrates the saturation of the city's road system.

As has been described, the poor quality of services, the drop in urban public transport users and, as a consequence, road congestion are problems in many Brazilian cities. These problems are often the result of a long period of lack of funding. In 2007, the announcement of the Growth Acceleration Programme (PAC) made official the resumption of investment in urban transport infrastructure in the country. Among the funding programmes created by the federal government, PAC 2 Mobilidade Médias Cidades (PAC 2 MMC) stands out, with investments for cities with 250,000 to 700,000 inhabitants (NTU, 2017), in which Joinville, with 577,077 inhabitants (IBGE estimate, 2016) was included and had its intervention proposals selected by the programme. Among the proposals submitted was the creation of exclusive corridors and lanes for urban public transport - buses, a characteristic element of a BRT system.

According to Branco (2013), the BRT system offers a fast, frequent, high-capacity, comfortable, less polluting service that is quickly implemented and has good network coverage, as well as requiring lower investment in implementation and operation. It is also considered flexible and can be easily adapted according to the demands of the market and the physical space where it will operate, thus justifying its implementation.

However, the lack of a single BRT concept has led to the construction of bus corridors that have been opened and named incorrectly. In order for these systems to be effective and provide a good user experience, as well as economic and environmental benefits, the Institute for Transportation & Development Policy (ITDP), in a global effort with transport leaders, created the BRT Quality Standard document in 2012. In order for a transport corridor to be identified as a BRT, it stipulates precise criteria that must be met (ITDP, 2016).

The need to solve urban problems, the potential of exclusive corridors, the existence of guiding documents and the federal government's investments in the municipality of Joinville through PAC 2 MMC became fundamental elements in choosing the topic. Castro (1997) points out three characteristics that must be met when choosing a research topic: importance, originality and feasibility.

Given the importance of documents that guide discussions about the BRT system, it is worth consulting them when working on a project to implement high-capacity bus corridors. This consultation makes it possible to identify the improvements or studies needed to achieve the efficiency of the public transport service on these routes.

The work is original in its approach because, in the course of the bibliographical research carried out, there were few studies at national level dealing with the operational project for implementing the BRT system, specifically in medium-sized cities as a single system and in large cities as a complement. In the state of Santa Catarina, there have been studies with analyses that have contributed to some approaches that lead to growth, as in the case of Joinville, a municipality that lacks investment in the current public transport system and has developed proposals for change. The projects included in the proposal to build bus corridors were therefore studied.

The case study is feasible because it combines the topic with the researcher's internship at SEPUD during the second half of 2017. The Secretariat showed interest in the work and provided all the necessary data to make it possible. In this way, the primary and secondary data were available to the researcher, as were the guiding documents for the analysis proposed in the work.

1.1. OBJECTIVES

1.1.1. General objective

To analyse the proposal for the construction of bus corridors made by Joinville City Hall and submitted to PAC 2 MMC, comparing it to the documents BRT Quality Standard 2016 and BRT Manual 2008.

1.1.2. Specific objectives

A) To study the exclusive bus lanes in the proposal submitted by Joinville City Hall to PAC 2 MMC;
B) Check that the system fits the classification of a BRT system according to the criteria set out in the BRT Quality Standard 2016 document;

C) Analyse whether the proposal complies with the guidelines set out in the BRT 2008 Manual document;

D) To verify the perception of urban public transport professionals in relation to the study's analysis.

1.2. WORK STRUCTURE

Figure 1 shows the structure of the work, which is made up of the following sections: introduction, theoretical framework, methodology, documents for analysis, case study, perception of professionals in the area, final considerations, references, annexes and appendix.

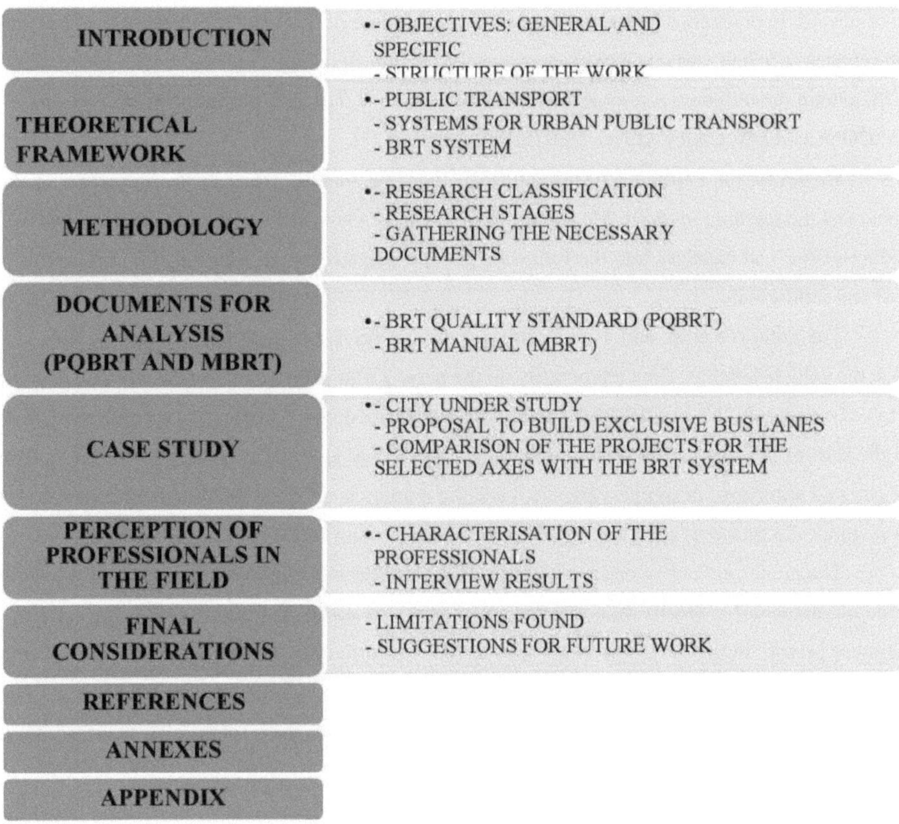

Figure 1 - Structure of the work.

Source: Author (2017).

CHAPTER 2

THEORETICAL FRAMEWORK

This section discusses public transport, urban public transport and its situation at national level. The different systems for urban transport are also presented, highlighting the BRT system.

2.1. PUBLIC TRANSPORT

For Preston (2009), public transport can be defined as any form of transport available for hire or reward. In practice, it generally refers to land passenger transport and, in particular, to bus and train services and their variants. "Public transport systems in Brazil are regulated by the government, which, among other duties, is responsible for setting fares for the services offered to the population" (VASCONCELLOS; CARVALHO; PEREIRA, 2011, pg. 27).

However, for Ferraz and Torres (2004), the **term 'public' refers to the ownership of the** vehicle and the freedom to use it. Modes of transport can be classified according to their ownership, freedom of use and capacity into the following groups: private or individual; public, collective or mass; and semi-public.

The authors Ferraz and Torres (2004) state that private or individual modes are those in which the vehicle belongs, even temporarily, to the person who is driving it, whether it is their own, borrowed or used. In this classification, there is complete freedom of choice as to the route to follow and the start of the journey, so there is flexibility in time and space to carry out the transport. The capacity and number of passengers are small and the journey is door-to-door. Common examples of private modes of transport are walking, cycling, motorbikes and cars.

The public, collective or mass classification characterises modes of transport used by many people simultaneously, which results in a low unit cost, and in which the vehicle belongs to a company or another person. In this way, the routes and timetables are fixed, i.e. there is no flexibility of use, and journeys are not door-to-door, with the need to complete them on foot or using other modes. The typical modes of urban public transport are: bus, metro, tram, aeromobile and suburban train (FERRAZ; TORRES, 2004).

According to Ferraz and Torres (2004), semi-public modes of transport are those that have intermediate characteristics between private and public modes. Examples include: taxis, buses (vans or minibuses transporting people with different levels of deregulation), charter buses, among others.

2.1.1. Urban public transport

The growth of urban centres makes it necessary to adapt urban development to transport and mobility needs, in search of mechanisms to preserve the environment and ensure social inclusion (COSTA et al., **2017). Reck (2017, p. 2) states that "urban evolution** has been, and to a certain extent always will be, conditioned to a scheme of circulation channels according to the transport technologies available to enable the necessary interrelationship between urban activities - residence, **work, study, leisure, etc".**

Private transport, unlike public transport, offers a free choice of route, travel at the most convenient time, in short, the versatility of door-to-door transport. However, public transport has the great advantage of saving public space, as it requires less than 10 per cent of road space compared to car transport. This is especially beneficial in central areas and places with limited space for circulation and parking (RECK, 2017).

Through the payment of a fare, or free of charge in some cases, public transport is accessible to the entire population and its services have the obligation to: permanently operate a specific transport network; transport users according to a fixed timetable; warn users in advance of the cost of services; and offer a special service to certain social groups or to serve certain regions. These legal or regulatory obligations are imposed by the public authorities and define and characterise public passenger transport services. These are represented by public transport (bus, train, metro) and individual transport (taxi) (RENK, 2017).

While public transport providers only operate on specific routes and serve predetermined areas, individual transport users benefit from a high quality of transport in terms of availability and time. However, the increase in individual car transport causes problems such as reduced road safety, increased air pollution, traffic noise and global warming. In view of these problems, it is essential to promote a change in people's behaviour by encouraging them to reduce car use by switching to public transport and the use of bicycles (NARAYANASWAMI, 2017).

According to Reck (2017), user behaviour should be treated more carefully in public transport, because in public transport the user acts passively, while in private transport they participate actively, and it is up to the public authorities to tackle the problems affecting public transport with greater emphasis and responsibility for solutions. To this end, the management of public transport requires more diverse and complex knowledge than in studies of individual transport, since it involves different modes of transport and requires direct intervention by the public authorities in its components (infrastructure, rolling stock, user behaviour and operating equipment).

2.1.2. Urban public transport in Brazil

In relation to the urbanisation process in Brazil, especially in medium-sized cities, the expansion of the urban territory has led to a dynamic of socio-spatial exclusion due to the formation of distant and discontinuous neighbourhoods, inhabited mainly by the poor. In this way, the lower-income population suffers in terms of getting around the urban environment due to its location, which leads to an increase in average travelling distances, making it difficult for the financially vulnerable to access the basic services that the municipality offers: health, education, leisure, among others (MARQUES; BRACARENSE, 2016).

Urban sprawl is a challenge for urban development. The ANTP (2006, pg. **82) points out that "although there is** a concentration of low-income populations in some deteriorated central areas, most Brazilian cities are spreading out over their territory, through the implementation of subdivisions, with low average occupancy". This dispersion of the population results in greater distances to be travelled to destinations offering jobs and services, thus increasing the costs of public transport and reducing its productivity. At the same time, it makes the use of individual transport more convenient and, in some cases, indispensable in the view of part of the population.

The Ministry of Cities (2004) reinforces this problem by pointing out that the difficulty in linking urban mobility to urban and regional planning is one of the main causes of the quality crisis in Brazilian cities. And that this contributes strongly to the generation of current scenarios where unsustainable cities can be seen from an environmental and economic point of view.

Currently, 82 per cent of the population represents urban Brazil. This urban concentration is explained when you realise that of the country's 5,561 municipalities, only 8%, or 455 municipalities, contain more than 55% of the country's total inhabitants. These include the 10 core cities of the most significant metropolitan regions. Supplying these regions, with increasingly distributed inhabitants, with urban facilities and public services, even if they fall short of basic needs, makes the overall costs of cities more expensive (MINISTÉRIO DAS CIDADES, 2004).

The fall in the use of public transport is a challenge for urban mobility. According to ANTP (2006), it is estimated that in recent decades there has been a drop of between 20% and 30% in the use of public transport in large Brazilian cities. More recent data presented by the National Association of Urban Transport Companies (NTU) shows that around 68 per cent of urban public transport companies have some debt and in three years - between 2014 and 2016 - the number of passengers fell by 16.5 per cent (CNT, 2017). This reduction increases environmental pollution and congestion, as most users switch to individual vehicles such as motorbikes and cars. The drop in users weakens the sector, jeopardising its structuring.

Funding is one of the challenges facing public transport in Brazil. According to ANTP

(2006, pg. 87),

> [...] there are two obstacles of a political and ideological nature, expressed by the priority given to individual transport over public transport. On the one hand, investments in the road system, which in most cases benefit car users, are defended as being in the public interest, while investments in large public transport systems become dependent on the financial market or the momentary (rare) availability of public funds. On the other hand, the costs imposed on society by car use are not properly accounted for, allowing socially regressive investments to be made.

Historically, the Brazilian transport system's main funding instruments for building roads, highways and public transport systems were the fuel consumption tax and the Single Road Tax (TRU), which were earmarked solely for this purpose. At the end of 1980, these sources of funding came to an end and the financing of infrastructure began to depend mainly on budgetary resources. As these became scarce, funding became increasingly restricted (ANTP, 2006).

According to ANTP (2006), after years of discussion, two permanent sources of funding were approved in Congress: the National Fund for Traffic Safety and Education (FUNSET), made up of 5 per cent of the value of all traffic fines collected in the country, and the Contribution for Intervention in the Economic Domain (CIDE), which comes from fuel consumption. However, both are not being used in practice. FUNSET has accumulated around R$600 million since the enactment of the new Traffic Code in 1998 until 2005, but the funds are being held in contingency by the Ministry of Finance. As for CIDE, the funds have been used to replace treasury resources for the railway sector and to pay off the sector's debts. Although 29% of the funds are earmarked for the states and municipalities, since 2004, not only has there been no formalised application to public transport, but most of the CIDE funds have also been contingent.

As far as operations are concerned, the first significant problem is the increase in operating costs per passenger due to the reduction in the number of users. Operating costs are paid for by user fares and, in some cases, such as on railways, mainly by budget top-ups (ANTP, 2006).

The second major problem is the increase in fares above the rate of inflation, caused, among other factors, by the increase in operating costs and gratuities. These are granted to certain user segments and passed on to paying users, leading to the exclusion of people on lower incomes or without access to transport vouchers (ANTP, 2006).

The lack of priority for public transport has negative impacts such as a reduction in speed and a drop in the regularity of service provision, reducing its attractiveness for users, who migrate to individual transport, usually cars and motorbikes. And, according to ANTP (2006, **pg. 89), "the amount of the** road **system** prioritised for public transport is minimal, almost **non-existent".**

According to Vasconcellos, Carvalho and Pereira (2011, pg. 68), **"in the case of buses, which** carry 90 per cent of public transport passengers, effective priority in circulation is minimal in most cities, generating very low speeds, around 30 per cent lower than would be the **case** with **adequate**

priority systems". Ensuring adequate operating conditions is therefore essential if we are to reverse the downward trends in reliability and competitiveness. Prioritising public transport can be achieved through a number of actions, such as the use of physical projects, for example: lanes, exclusive channels, corridors and/or electronic control of the use of road space, combined with restrictions on car use in situations where this is essential.

Finally, the Ministry of Cities (2004, pg. 6) points to two aspects that contribute to the mobility crisis that is **evident in urban centres: "the increase in** externalities produced by individual transport - accidents, congestion and pollution; and social exclusion - due to the inability of the poorest part of the population to pay **fares".** In this way, the way in which urban public transport is currently planned and operated induces, not always voluntarily, the disorderly occupation of land.

In other words, even though there is little public control, transport remains a determining factor in the possibility of occupying and consolidating new spaces. In this way, it is easy to recognise that by including urban transport in the integrated planning of municipalities, in a way that incorporates the principles of full sustainability and has its planning and operation geared towards the interests of the population, it can become an important and effective instrument for structuring and vectoring controlled urban expansion. (MINISTÉRIO DAS CIDADES, 2004).

2.2. SYSTEMS FOR URBAN PUBLIC TRANSPORT

Due to the different urban dynamics and characteristics of each region, public transport must adapt and be constantly adaptable, both in terms of numbers **and the technology used. In this way, "the various modes that make it up** (metro, trolleybus, buses, etc.) must function as a Public Passenger Transport System (PPTS), each serving specific functions, in the constant search for **operational rationalisation and a reduction in transport costs"** (RECK, 2017, pg.4).

With regard to the characteristics of the regions, the authors Ferraz and Torres (2004) state that, to a large extent, the size of the city determines how its inhabitants get around. In the case of very small cities, travelling is almost exclusively done on foot. As the size of the city increases, so does the use of individual transport (private transport by car, motorbike and bicycle and/or semi-public transport by taxis).

In medium-sized cities, on the other hand, there are wider roads, making public transport by bus possible. However, in larger cities, expressways appear, especially on the banks of rivers, and there is often a prioritisation of public transport through the use of exclusive lanes for buses and trams (FERRAZ; TORRES, 2004).

In large cities, pre-metro public transport systems, viaducts and expressways appear. In metropolises it is common for the main public transport systems to comprise metro, pre-metro and articulated or bi-articulated buses in segregated lanes (FERRAZ; TORRES, 2004).

According to Reck (2017), due to the current economic reality, solutions that require high investment, such as the metro, train and trams, have been restricted to large metropolises, leaving other cities to solve their transport problems through non-capital-intensive solutions.

The authors Reck (2017) and Ferraz and Torres (2004) highlighted the following modes of public transport: buses, trolleybuses, trams, metro and pre-metro. However, the author Andrade (2015) points to a more extensive list of modes for urban public transport: buses with internal combustion engines, electric buses, BRT, LRV, Light Rail or Segregated LRV, Monorail, Metro and Metropolitan Train.

Sousa (2014) points out that the word bus comes from Latin (omnibus) and means **"for all".** The author also states that the bus is the most common mode of public transport in Brazil. Limited investment partly justifies the predominance of bus systems in many Brazilian capitals and large cities.

Buses are made up of different types of vehicles, which are called Standard, Articulated and Biarticulated, characterising different sizes and capacities. Buses run on public roads and share them with cars, cyclists and pedestrians (AFFONSO NETO, 2013). Trolleybuses, on the other hand, are electric buses that rely on an aerial network of wires that connect the vehicle to electricity (LOBO, 2014).

The BRT system differs from a conventional bus system for a number of reasons, including: the use of a segregated, exclusive lane; ensuring level boarding and alighting on the platform; high commercial speed; ensuring advance payment of the fare and providing information to users via the operational control centre. These elements are essential for a bus corridor to be classified as a legitimate BRT, according to Castro (2013). In Brazil, the cities of Curitiba (PR), Cuiabá (MT), Goiânia (GO), Brasília (DF), Belo Horizonte (MG), Uberlândia (MG), Rio de Janeiro (RJ), São Paulo (SP), Uberaba (MG), Recife (PE) and Aracaju (SE) have BRT corridors (BRT BRASIL, 2016).

As for the use of bus lanes, there are two classifications for the type of lane used, characterising different transport systems. The first classification concerns the BRT, as seen above, which uses segregated lanes, i.e. exclusive corridors in which public transport is not interfered with in any way by private vehicle traffic. There are, however, preferential lanes for public transport, but they allow other vehicles to pass when there is a need to make an immediate right turn or access neighbouring plots. The second concerns the **Bus Rapid System** (BRS or Bus Rapid Service), **which "includes, in** addition to the implementation of preferential lanes for public transport vehicles, the optimisation of the service offer (by reducing the bus fleet in the corridor and increasing its occupancy), the staggering of bus stops, as well as the control of **private vehicle access by cameras and/or sensors"** (RODRIGUES, FILHO E GOMES, 2013, pg. 23).

The author Affonso Neto (2013) points out the characteristics of the following guided vehicles: Tram, Light Rail Vehicle (LRV), Monorail, Light Metro, Urban Metro (heavy) and Regional Metro (Metropolitan Train). The tram is a vehicle on rails that circulates on public roads, sharing them with mixed traffic. The Tram serves the existing transport offer between buses and heavy metro and its capacity differs in terms of the technology used and the level of segregation. The Monorail requires little space to lay tracks and uses the same signalling and control systems as the Metro. The Light Rail has vehicles with a reduced gauge, with a smaller height, width and length than the classic metros. The Metro is characterised by its total segregation and high degree of automation. The Metropolitan Train also has total segregation, with the distance between stations being greater than 1,500 metres, and, in general, the vehicle runs on the surface but uses underground sections in the central zones. In order to observe more specific characteristics of some of the modes mentioned, Table 1 shows the main technical parameters of urban public transport modes.

Table 1 - Typical ranges of values for the main technical parameters of urban public transport modes

Parameters	**Buses**	**Tram**	**VLT**	**Metro**	**Suburban train**
Width (m)	2,4 - 2,6	2,4 - 2,6	2,4 - 2,8	2,5 - 3,2	2,5 - 3,2
Unit length (m)	6,5 - 24	14 - 23	14 - 30	15 - 23	20 - 26
Capacity (pass/unit)	20 - 240	70 - 250	100 - 250	150 - 250	150 - 250
Units in convoy	1 - 4	1 - 3	1 - 4	4 - 10	4 - 10
	Independent	Engaged	Engaged	Engaged	Engaged
Speed (km/h)	10 - 40	10 - 20	20 - 40	25 - 60	40 - 70
Transport capacity (thousand pass/h)	5 - 15	5 - 15	10 - 25	25 - 60	20 - 50
Distance between *stops (m)*	200 - 400	200 - 400	400 - 800	700 - 2000	1500 - 4000

Source: Rodrigues, Filho and Gomes (2014) apud Ferraz (2004).

Table 1 shows that the bus, the predominant mode of transport in Brazilian cities, has the lowest capacity and ties with the tram in terms of transport capacity. Speed is also one of the lowest, thus showing the inferiority of the conventional bus system in relation to the other modes presented, if these aspects are analysed.

As the bus is the predominant transport system in Brazil, as highlighted by Sousa (2014), it is worth presenting Figure 2. This shows the conventional bus system, with different vehicles and different structures, the BRT and the guided modes VLT, Light Metro and Heavy Metro.

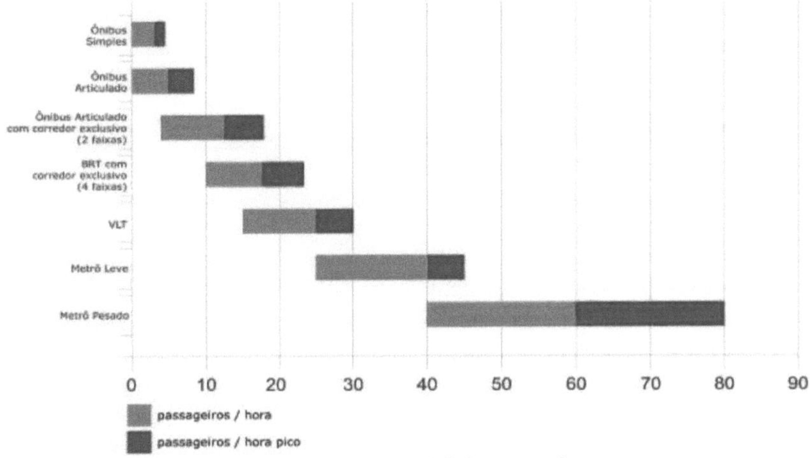

Figure 2 - Transport capacity of different modes.

Source: Castro (2013) apud Alouche (2013).

Figure 2 shows the importance of exclusive corridors in the capacity of the transport system and shows that the capacity of the BRT system is close to that of a tram system. Therefore, given the predominance of buses as a mode of public transport and the importance of exclusive corridors for increasing capacity in the system, the BRT system is highlighted in the next section.

2.2.1 Bus Rapid Transit (BRT) system

The BRT Manual (MBRT) document developed by the Institute for Transport and Development Policy (ITDP) and translated by the Ministry of Cities (2008, pg. 1) defines BRT as "a bus transport system that provides fast, comfortable and cost-efficient urban mobility through the provision of segregated infrastructure with transit priority, fast and frequent operation and excellence in marketing and **user** service".

The Ministry of Cities (2008) points out that BRT has performance and comfort characteristics very similar to rail transport systems, but at a fraction of the cost of implementation. Compared to a tram or tram system, a BRT system costs 4 to 20 times less and compared to a metro system the cost is 10 to 100 times less.

In turn, the BRT Quality Standard (PQBRT) document developed by ITDP (2016, p. 4), in partnership with different transport leaders, points out that **BRT are** "medium and high capacity public transport systems, based on exclusive bus lanes, which provide fast, comfortable and **cost-effective** transport". **For** the system to achieve high performance, superior quality of service and a better user experience, it must combine five fundamental characteristics that define a BRT system: segregated infrastructure with exclusive circulation; alignment of bus lanes; fare collection outside the bus; level boarding and treatment of intersections with priority passing.

It's worth noting that the 2014 World Cup and the 2016 Olympics motivated the federal government to implement the Growth Acceleration Programmes, the Urban Mobility PAC and the Big Cities Mobility PAC. These investments have enabled Brazilian cities to pay more attention to urban mobility and public transport (BRT BRASIL, 2016). This led to the creation of the BRT Brasil programme, developed by the NTU (National Association of Urban Transport Companies), which aims to monitor the implementation of bus rapid transit systems.

"[...] the BRT system is the most complete and efficient mobility and bus transport solution for medium and large cities, and has been the biggest beneficiary of **these investments."** From the start of investments until May 2016, 19 states and 33 cities totalled 101 projects **with a length of 1,472.5 km.** "[...] of these projects, 22 are in operation, 21 are under construction and 58 are in the design phase, prioritising **urban bus transport."** The municipalities of Cuiabá (MT), Goiânia (GO), Brasília (DF), Belo Horizonte (MG), Uberlândia (MG), Curitiba (PR), Rio de Janeiro (RJ), São Paulo (SP), Uberaba (MG), Recife (PE) and Aracaju (SE) have BRT corridors (BRT BRASIL, 2016).

CHAPTER 3

METHODOLOGY

This chapter presents the classification of the research as well as its stages and the main documents used to discuss the subject of this work.

2.3. RESEARCH CLASSIFICATION

According to Silva and Menezes (2005), research has various definitions, but simply put, it means seeking answers to proposed questions. In addition, research is a set of actions designed to solve a problem, based on rational and systematic procedures. Research is carried out when there is an unsolved problem.

From the point of view of the nature of this work, the research is applied, i.e. it seeks to generate knowledge and then apply it to specific problems. In terms of approach, this is qualitative research, as it focuses on the relationship between the researcher and the study environment, where the environment is the direct source for data collection and the researcher is the key element. This data cannot be translated into numbers, but is given meaning as a result of the researcher's intuitive analysis.

From the point of view of the objectives of the work, the research can be classified as exploratory and descriptive. The research is exploratory due to the need to familiarise oneself with the subject, through bibliographical and documentary research. It is also characterised as descriptive, as it describes the characteristics of the environment and the operations involving data collection.

In terms of technical procedures, the research is characterised as bibliographical, documentary and case study. As described above in terms of objectives, bibliographical research is necessary to generate knowledge about the subject. It is also classified as documentary because of the need to manipulate data that has not yet been studied. And it is a case study because, as in the definition of applied research, it generates knowledge and applies it to a specific problem.

3.2. RESEARCH STAGES

This section briefly describes the research stages. These are shown in Figure 3 and characterise the processes of this work.

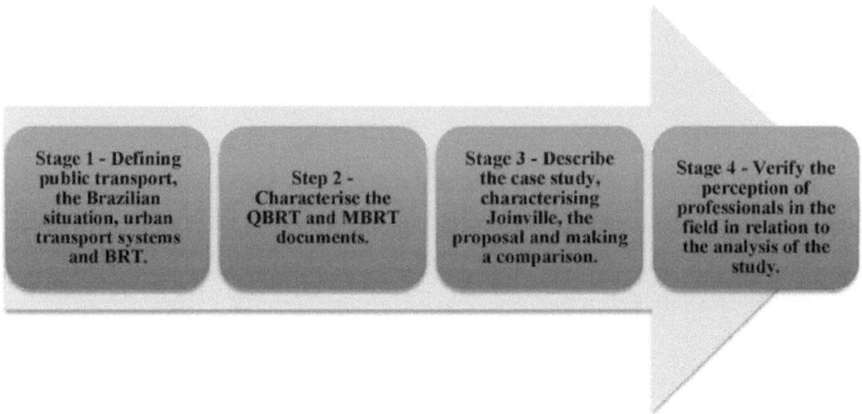

Figure 3 - Research stages.

Source: Author (2017).

In the first stage, public transport is defined, covering not only its definition in itself, but also characterising urban public transport and the situation in Brazilian cities. This stage also defines urban transport systems, presenting different modes of transport, characterising them and highlighting their differences, with emphasis on passenger capacity. Finally, the BRT system and its characteristic elements are defined.

The second stage characterises the PQBRT (BRT Quality Standard) and MBRT (BRT Manual) documents, presenting their origins and objectives. The items discussed throughout the case study are also highlighted.

The third stage presents the case study. It introduces the municipality where the study is being carried out, presents the proposal for building exclusive corridors and compares it to the documents characterised in the second stage.

The fourth and final stage seeks to verify the perception of urban public transport professionals in relation to the analysis of the study by means of an interview. The professionals are characterised and the operational planning of Joinville's Integrated Transport System (SIT) is discussed.

To develop the interview script, the central elements of the instrument were the problems observed during the analysis of the case study, in order to reinforce points and clarify the participation of SIT's operational planning sector in the preparation of public transport projects developed by Joinville City Hall. The script drawn up for the interview can be found at the end of this work as an Appendix.

The interviewees were chosen according to their area of activity, at Passebus, the company responsible for ticketing, database administration, customer service and Citizenship Stations, which work directly with the operational planning of the Joinville SIT. Sampling is non-probabilistic and intentional.

3.3. COLLECTING THE NECESSARY DOCUMENTS

i. BRT Quality Standard Document, 2016 edition, ITPD (**Institute for Transportation & Development Policy).**
ii. BRT Bus Rapid Transit Planning Guide, 2008 edition, published by **ITPD (Institute for Transportation & Development Policy)** and the Ministry of Cities.
iii. Joinville City in Data, 2017 edition, published by SEPUD (Urban Planning and Sustainable Development Department), Joinville City Hall.
iv. QualiÔnibus Joinville Satisfaction Survey, 2014 edition, organised by EMBARQ Brasil[1] .
v. Preliminary booklet: Joinville Sustainable Mobility Plan (PlanMOB), 2015 edition, IPPUJ (Foundation for Research and Planning for the Sustainable Development of Joinville[2]), Joinville City Council.
vi. Diagnosis part 1: Joinville Mobility and Accessibility Plan (PlanMOB), 2014 edition, IPPUJ (Foundation for Research and Planning for the Sustainable Development of Joinville), Joinville City Council.

[1] Current WRI Brasil Sustainable Cities.
[2] Current Secretary of Urban Planning and Sustainable Development (SEPUD).

CHAPTER 4

DOCUMENTS FOR ANALYSIS (PQBRT AND MBRT)

The BRT Quality Standard documents (2016) and the BRT Manual (Planning Guide, 2008) were used in the case study as the basis for the analyses in sections 5.3.1 and 5.3.2. It is therefore necessary to describe them for a better understanding.

4.1. BRT QUALITY STANDARD (PQBRT)

The lack of alignment between planners and engineers when implementing a new high-quality BRT corridor simultaneously led to dozens of other corridors being incorrectly named as BRT, because a few years ago there was no single understanding of the BRT concept. With this in mind, the ITDP (**Institute for Transportation & Development Policy),** in a global effort with transport leaders, created the PQBRT document in 2012 and its latest version was published in 2016.

The PQBRT document makes it possible to evaluate BRT systems on the basis of international best practice and acts as a planning tool. It defines the essential elements of a BRT system and provides a means of arriving at a common definition of the system. The document provides a framework that allows system planners, managers and the sustainable transport community to identify and implement quality BRT corridors, thus ensuring that these systems can offer a uniform and superior experience to their users, as well as significant economic benefits and positive environmental impacts.

According to ITDP (2016), a corridor can be certified as a gold, silver, bronze or basic BRT, scoring it through categories and metrics related to both the design and operation of the corridor. This positions it as one of the best international practices. The cities in which BRT corridors are certified take urban public transport to a new level of excellence and become examples of good practice. In this way, by clarifying and raising the standards under which all BRT systems are built through the PQBRT document, more people will experience the convenience and comfort of this mode of transport, and more municipalities will be favoured by an efficient, cost-effective medium- and high-capacity transport system.

ITDP hopes that by helping to define and recognise good quality BRT systems, this will bring about the fundamental and necessary change to prioritise public transport. To this end, the PQBRT document emphasises operational and safety aspects to ensure that corridors with a good rating, according to the criteria described in the document, continue to offer a quality service and an excellent travel experience to users (ITDP, 2016).

The BRT systems awarded gold were the BRT of Yichang (China), MOVE of Belo Horizontes (Brazil) and TransMilenio of Bogotá (Colombia). The silver winners were Metrobus in Buenos Aires (Argentina), Metrobús in Mexico City (Mexico), Rea Vaya in Johannesburg (South Africa) and CTfastrak in Hartford (United States). And only the Rainbow BRT in Primpri-chinchwad (India) was awarded bronze. Table 2 shows the scoring system defined in the PQBRT document (ITDP, 2016).

Table 2 - Scoring system.

Category	Maximum score
Basic BRT (pages 26 - 37)	38
Segregated infrastructure with priority passage	8
Alignment of bus lanes	8
Fare collection outside the bus	8
Treatment of intersections	7
Level boarding	7
Service Planning (p. 38 - 44)	19
Multiple lines	4
Express, limited and local services	3
Control centre	3
Location in the top ten corridors	2
Demand profile	3
Hours of operation	2
Multi-corridor network	2
Infrastructure (p. 45 - 52)	13
Overtaking lanes at stations	3
Minimising bus emissions	3
Stations away from intersections	3
Central stations	2
Floor quality	2
Stations (p. 53 - 57)	10
Distance between stations	2
Safe and comfortable stations	3
Number of bus doors	3
Shoulder bays and stop sub-points	1
Sliding doors at BRT stations	1
Communications (p. 58 - 59)	5
Brand consolidation	3
Passenger information	2
Access and Integration (p. 60 - 65)	15
Universal access	3
Integration with other modes of public transport	3
Road safety and pedestrian access	4
Secure bicycle parking	2
Cycle infrastructure	2
Integration with bike-sharing systems	1
Operating Deductions (p. 66 -72)	-63
Commercial speeds	-10
Peak passengers per hour per direction (pphps) less than 1,000	-5
Lack of enforcement of priority of passage	-5
Considerable gap between the bus and the platform	-5

Overcrowding	-5
Poor infrastructure maintenance	-14
Low peak frequency	-3
Low off-peak frequency	-2
Unsafe use of bicycles	-2
Lack of road safety data	-2
Bus routes parallel to the BRT corridor	-6
Bus convoy formation	-4

Source: Author (2017).

The PQBRT document should not be applied to the entire BRT system, but rather to specific corridors. This is justified when analysing its quality in cities with multiple corridors in operation, which can vary significantly. A BRT corridor is defined in the PQBRT document, ITDP (2016, pg. 26), as "a section of a road or contiguous roads served by one or multiple bus lines, with a minimum length of 3 kilometres, with a segregated bus-only lane"

The QBRT document justifies the way the corridor is defined by stating that in some cities, BRT does not have priority over car traffic and this is an essential element in improving both its efficiency and cost. To avoid systems that don't use this policy, the corridor should be defined to include segregated lanes exclusively for buses. This definition does not include corridors that allow other traffic, as the presence of other vehicles can reduce the speed of buses and the regularity of the service offered to users, thus resulting in a decrease in the attractiveness of the system compared to the private individual (ITDP, 2016).

According to ITDP (2016), Basic BRT, the first category in Table 2, is a set of elements considered by the Technical Committee to be essential for defining a corridor as BRT. This set is made up of five elements that contribute to eliminating delay factors caused by congestion, conflicts with mixed traffic and passenger boarding and alighting, thereby increasing efficiency and lowering the system's operating costs. These are: segregated infrastructure with transit priority, bus lane alignment, off-bus fare collection, intersection treatment and level boarding. It is worth highlighting Tables 1, 2, 3, 4 and 5, which show the Basic BRT score, as these are used for comparisons in section 5.3.1.

The scoring guidelines in Table 1 state that the score is calculated by multiplying "the percentage of the corridor that has each type of segregated infrastructure with priority passage for BRT services X the number of points associated with the type of segregation". Corridors that allow other vehicles such as taxis and vans to enter do not receive a score, as they are not considered segregated infrastructure (ITDP, 2016, pg. 28).

Table 1 - Type of infrastructure with priority for passage.

Type of segregated infrastructure with priority passage	POINTS	PONDERING
Segregated and physically separated lanes.	8	
Segregated lanes of a different colour, with no physical separation.	6	

Segregated lanes separated by a line painted on the pavement.	4	% of the corridor with the type of segregation with priority of passage
No segregated lanes.	0	

Source: ITDP (2016).

The scoring guidelines presented in Table 2 define that the score "is weighted, using the percentage of the corridor in each particular configuration, multiplying it by the number of points associated with that configuration and then adding these numbers together" (ITDP, 2016, pg. 29).

Chart 2 - Corridor configurations.

Corridor configurations	POINTS	PONDERING
LEVEL 1 SETTINGS		% of corridor with 0 type of bus lane alignment
Bus lanes in both directions aligned on the centre lane of a two-way street.	8	
Bus-only corridors with full transit priority and no parallel mixed traffic, such as transit malls (e.g. Bogotá, Colombia; Curitiba, Brazil; and Quito, Ecuador) and converted rail corridors (e.g. Cape Town, South Africa; and Los Angeles, USA).	8	
Bus lanes that run adjacent to the shores of lakes, rivers or parks, where there are few intersections that could cause conflicts.	8	
Two-way bus lanes on the side of a one-way street.	6	
LEVEL 2 SETTINGS		
Bus lanes divided into pairs of one-way lanes on separate streets, with each bus lane aligned centrally on the road.	5	
Bus lanes aligned with the outside kerb of the centre lane on a road with a centre lane and a side service lane.	4	
Bus lane aligned with the inner kerb of the service lane on a road with a centre lane and side service lane. The bus lane must be physically separated from other traffic on the service lane in order to receive points.	4	
Bus lanes divided into pairs of one-way lanes on separate streets, with each bus lane aligned with the pavement.	3	
LEVEL 3 SETTINGS		
Virtual bus lanes that operate in both directions on a single median and alternate 0 directions per block.	1	
CONFIGURATIONS WITHOUT PUNCTUATION		
Bus lane aligned with the pavement in a two-way street.	0	

Source: ITDP (2016).

Table 3 shows the scoring guidelines, which **state that** "to receive a score, off-bus fare collection must take place during all operating hours. The score is weighted by the percentage of stations or lines in the corridor that use the payment system. The maximum score for this element is eight **points"** (ITDP, 2016, p. 35).

Chart 3 - Fare collection outside the bus.

Fare collection outside the bus (during all operating hours)	POINTS	PONDERING
Access control	B	% of stations in the corridor
Proof of payment	7	% of lines using bus corridor infrastructure
Ticket validation on board - all doors	4	% of lines using bus corridor infrastructure

Source: ITDP (2016).

Table 4 shows the guidelines which **state that** "scores are weighted by the percentage of prohibited turns or intersections with traffic light priority along the corridor." The total score is the sum of the points obtained for the different situations that exist. When the corridor is not at the same level as mixed traffic, the conversion is considered prohibited. And finally, the score limit is seven even if it is the sum of a value above it (ITDP, 2016, pg. 36).

Table 4 - Treatment of intersections.

Intersection treatment	POINTS	PONDERING
Prohibited conversions in the bus lane.	7	% of prohibited turns on the bus lane
Traffic light priority at intersections.	2	% of corridor intersections

Source: ITDP (2016).

With regard to level boarding, buses that have a vertical span, the distance between the vehicle floor and the boarding platform, of more than four centimetres are not **classified as "level with platform". If the bus has steps inside, it is also not considered "level with platform". In** addition, **according to the guidelines, "the** scores for each element are weighted by the percentage of buses that are level with the platform and the percentage of stations that adopt measures to reduce the horizontal span." And just like the previous element, the limit is seven points (ITDP, 2016).

Chart 5 - Level boarding.

Level boarding	POINTS	AUTHORISATION
The buses are level with the platform, with a vertical gap of no more than 4 centimetres.	7	% of buses operating on the corridor
The stations in the corridor have measures to reduce the horizontal span.	6	% of stations in the corridor

Source: ITDP (2016).

Finally, the PQBRT document defines the basic requirements for a system to be considered a BRT as the need to have at least three kilometres of segregated bus-only lanes, to score at least four points in the "segregated infrastructure **with priority of passage" element** (Table 1), to score at least four **points in the "alignment of bus lanes" element** (Table 2) and for the final score to be at least 20 points in the total of the five Basic BRT elements (ITDP, 2016).

4.2. BRT MANUAL (MBRT)

The MBRT document was developed by the ITDP (**Institute for Transportation & Development Policy)** and translated by the Ministry of Cities (2008), which details the steps for realising a successful BRT system and divides them into five major planning areas. These areas include project preparation, operational design, physical design, integration, the implementation plan and evaluation and implementation.

For the analyses carried out in the case study, section 5.3.2, the chapters referring to the operational project were used. It is therefore worth describing the chapters covered by the Manual. The operational project covers four chapters and is discussed in the following sections of this work: 4.2.1 Network and line design, 4.2.2 System capacity and speed, 4.2.3 Intersections and traffic light control and 4.2.4 User service.

4.2.1. Network and line design

The MBRT document points out that the starting point for a project should be neither the infrastructure nor the vehicles, as the system should be designed to achieve the operational characteristics that are expected for the user. From the user's point of view, the most important factors in selecting a mode of transport are whether it serves the places they want to go and how long the journey takes.

The network and line design chapter of the MBRT document points out the definition, advantages and disadvantages between open and closed systems, because, according to the **Manual, "the extent of access** given *to* predetermined operators and vehicles directly affects vehicle speeds, environmental impact and the aesthetic qualities of the **system". It also discusses** trunk **services versus** direct services, and in **a designed system it is possible to "accommodate a range of population densities to achieve an urban-wide service".** Also, with regard to **line** design, **"a** well-designed system can optimise journey times and convenience for **a greater number of journeys and significantly reduce operating costs"** (MINISTÉRIO DAS CIDADES, 2008, pg. 231, pg. 235).

4.2.2. System capacity and speed

According to the MBRT document, capacity, speed and frequency of service are the characteristics that differentiate BRT systems from conventional services and buses. Designing a BRT system that can comfortably handle high user demand and minimise travel and waiting times is one of the pillars for promoting a service that can compete with cars.

The operational design chapter of the MBRT document deals with decisions that affect the basic parameters of capacity, speed and frequency. An important step for efficient design is to identify all the critical elements that can inhibit high-capacity and high-speed service, because as the number of passengers and vehicles increases, so do the chances of bottlenecks and operational problems.

The topics discussed in this chapter of the MBRT document are: "Calculating required capacities"; "Vehicle size"; "Station-vehicle interface"; "Multiple stop bays and express services"; "Trains"; and "Station spacing". These outline the characteristics and design that can enable a system to achieve both high capacity and high speed.

The topic "Calculation of required capacities" points out the importance of emanate analysis and the milling process to help quantify existing public transport demand, as well as providing projections of expected system growth. The topic "Vehicle size" states that the capacity of the selected bus determines the number of vehicles for the corridor, thus presenting vehicle-related calculations such as the required fleet. In the topic "Station-vehicle interface", **innovations such as** level boarding, charging outside the bus, efficient bus alignment in the station, multiple wide doors, and comfortable space in the stations for users make boarding and alighting faster, thus reducing congestion in the stations and therefore offering higher-capacity services. The topic "Multiple stop bays and express services" **gives examples of implementing** more stop bays in order to increase the system's capacity. The topic "Trains" describes a train system that allows multiple stopping positions without an overtaking lane. And finally, the topic "Station spacing" **deals with the influence of the distance** between stations on the capacity and speed of the system (MINISTÉRIO DAS CIDADES, 2008).

4.2.3. Intersections and traffic light control

The MBRT document points out that signalised intersections are critical points along any BRT corridor and therefore the capacity of the system can be significantly reduced due to a poorly designed intersection or poorly designed signal stages. Finding solutions to optimise the performance of an intersection can make a big difference to improving the efficiency of the system.

The introduction of a poorly designed BRT system can result in problems at intersections such as a multiplication of traffic light stages and/or a multiplication of turning lanes, interfering with both bus and mixed traffic flows and consuming crossing areas at the turn that could be dedicated to pedestrian improvements or even avoiding costly land acquisition. The MBRT document points out that, in general, "the aim of designing an intersection for a BRT system is to minimise waiting, improve the safety and convenience of access to bus stations for pedestrians and minimise waiting for mixed traffic" (MINISTÉRIO DAS CIDADES, 2008, pg. 300).

According to the MBRT document, to avoid the aforementioned problems and achieve the objective described above, BRT system planners approach intersections in the following ways: by identifying existing bottlenecks and resolving them using normal engineering practices; by simplifying the line structure of the BRT system; by calculating traffic light delays for the new BRT system; by restricting turning movements in BRT corridors as much as possible; by making decisions about the approach to turning movements for vehicles within the BRT system; by optimising station locations, by optimising intersection design and traffic light stage times; by considering traffic light priority for public transport vehicles in low-volume systems; and, in high-volume systems, by considering bus lane surface separations at intersections.

These approaches are discussed in this MBRT document chapter through the **following topics: "Intersection evaluation"; "Conversion movement restriction"; "BRT conversion designs"; "Station location in relation to the intersection"; "Roundabouts"; and "Traffic light priority". These provide some generic rules to help with** project decision-making in the cases mentioned above.

4.2.4. User service

The MBRT document highlights the importance of the user for the BRT system by stating that "the quality of the service to the user is directly related to user satisfaction, which will ultimately determine utilisation and long-term financial sustainability" (MINISTÉRIO DAS CIDADES, 2008, p. 340). Thus, unlike many conventional bus systems in developing cities, in a BRT system the needs of the user are placed at the centre of the project and its implementation criteria.

This chapter includes aspects of user information, system professionals, safety, comfort facilities and division of services. With regard to user information, the MBRT document, Ministry of Cities (2008, pg. 341), highlights the need **for clear maps of the system, because "the lack of clarity of the system is a formidable barrier,** especially for new users and those residents with only **occasional** transport needs**".** Station **signage** is essential for the user to understand the system, providing information such as platform and line identification, directions for entering, exiting or asking questions, ticketing instructions, among other information needed to fully understand the system. Also covered are public service messages, visual and voice information systems, user interaction, call centres, email contacts and information websites, user service desks, survey forms and public representation.

As for the system's professionals, the MBRT document emphasises the role of public transport staff as one of the best promotional tools available, who must make users feel respected and **welcome. Because, "as in life,** sometimes a simple smile or a kind word **can make all the difference"** (MINISTÉRIO DAS CIDADES, 2008, pg. 351). The subjects of cleanliness, hygiene and the provision of drinks and food at the stations are also covered.

For user safety, the MBRT document points to the need for driver training, regular vehicle inspections and written maintenance procedures as the basic elements of a safety programme. The MBRT document not only addresses traffic safety, but also public safety, as the fear of crime and robbery is a major motivating factor in the migration to individual transport.

The topic of comfort facilities highlights that a major advantage of public transport over individual transport is the time that can be used for other activities such as reading, chatting with friends and relaxing. The MBRT document, Ministry of Cities (2008, pg. 359), **states that** "comfort facilities can help make the most efficient use of this **value-added time".**

The division of services, in turn, discusses the advantages and disadvantages of a BRT system serving and offering different services according to the type of user. Where the advantage lies in reaching users who would otherwise not use public transport and the disadvantage of this segmentation is the possible increase in complexity in managing this system.

CHAPTER 5

CASE STUDY

This section discusses the city of Joinville, the subject of the case study, and the proposal made by the city council to build bus corridors in the city. This is referred to throughout the paper simply **as a 'proposal'.** The axes selected for study and the population's demands regarding public transport in the region are highlighted. The proposal for the construction of bus corridors made by Joinville City Hall and submitted to PAC 2 MMC is also analysed, comparing it to the MBRT and PQBRT documents.

4.3. JOINVILLE

Joinville is situated in the north-eastern micro-region of the state of Santa Catarina, located in the southern region of the country, and is the largest city in Santa Catarina. According to SEPUD[3] (2017a), in 2016 the city ranked 37ª among the largest exporting municipalities in Brazil and 2nd in the state. Currently, the municipality is responsible for around 20 per cent of Brazil's exports.

The city is also an industrial centre in the southern region, as much of its economic activity is concentrated in industry. The main industrial sectors in Joinville's economy are metalworking, textiles, plastics, metallurgy, chemicals and pharmaceuticals. The volume of revenue generated for the public coffers by the municipality is second only to the capitals Porto Alegre (Rio Grande do Sul) and Curitiba (Paraná), considering the southern region of the country. Joinville's Gross Domestic Product (GDP) also stands out, ranking 28th in the country (SEPUD, 2017a).

The city's growth is directly linked to the expansion of industry, which brought with it population growth. The city's economic profile began to broaden in the 1990s, encompassing the service and technology sectors, through the decentralised commercial development of the neighbourhoods, which are increasingly independent of the city centre, and simultaneously with the population growth rate, which stabilises and remains at around 1.50% per year. The city's first large shopping malls opened in the mid-1990s and, with globalisation, the region's largest companies consolidated their national and international leadership (SEPUD, 2017a).

3SEPUD - Joinville Urban Planning and Sustainable Development Department. Joinville City in Data 2017. Joinville: City Hall, 2017a, 73f.

According to the IPPUJ[4] (2015a), Joinville has 528,476 inhabitants in 210 square kilometres of urban area and a density of 2,514 inhabitants per square kilometre. The urban area is highlighted in Figure 4 in dark grey. It can be seen that, despite the predominance of rural areas over urban areas, the significant urban density prioritises the management of urban areas.

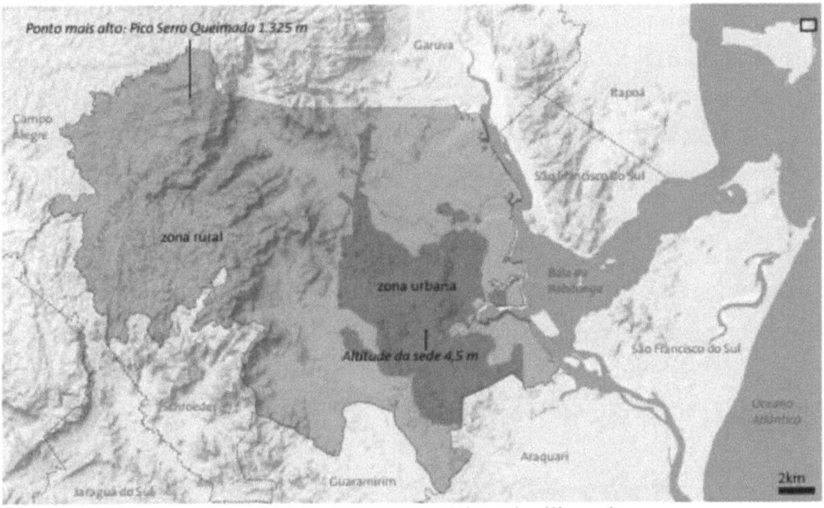

Figure 4 - Map of the Joinville region.

Source: Joinville City Hall (2017).

As for mobility in the urban area, the most important issues are walkability, the cycle network and public transport. The IPPUJ (2015a) states that the pavements have serious problems with accessibility, continuity of routes and lack of standardisation, so they are not conducive to pedestrian movement. The cycling network needs to be expanded, but the continuity of its route should also be prioritised, integrating the existing with the proposed.

The city currently has 125 kilometres of cycle paths, which could be extended in order to take up the city's tradition where, in the mid-1970s, cycling made up 30% of all journeys.

With regard to public transport, IPPUJ (2015a) points out that the municipality has around 14 kilometres of preferential lanes, which may not be enough for a city of its size and demand. Among the improvements needed, 20% of the city's fleet still needs to be made accessible, and several bus shelters need to be refurbished to provide protection against bad weather, greater comfort and information for users, increasing the attractiveness of the system. In the same way, the integration terminals need to be restructured so that they effectively become hubs for commuting, commerce and

[4] IPPUJ - Fundação Instituto de Pesquisa e Planejamento para o Desenvolvimento Sustentável de Joinville (Org.). **Preliminary Notebook: Joinville Sustainable Mobility Plan (PlanMOB)**. Joinville: City Hall, 2015a, 119 f.

services, and a reference point for neighbourhood centres.

Finally, the IPPUJ (2015a) states that Joinville has no mechanisms to regulate or condition car transport, thus contributing to the inefficiency of the entire mobility system. To this end, it was necessary to think about strengthening public management, prioritising collective and non-motorised transport, social inclusion, democratic management and environmental sustainability.

Having highlighted the need to prioritise public transport, the importance of projects geared towards it becomes clear. That said, section 5.2 presents the proposal made by Joinville City Hall to PAC 2 MMC for the construction of exclusive bus lanes.

4.4. PROPOSAL TO BUILD EXCLUSIVE BUS LANES

This section describes the proposal to build exclusive bus lanes drawn up by Joinville City Hall to improve urban mobility in the municipality. This is referred to throughout the **paper simply as a 'proposal'.** In this way, the PAC 2 MMC is conceptualised, the axis selected for study is defined and the demands of the Joinville population are highlighted.

4.4.1. Growth Acceleration Programme 2 (PAC) Medium Cities Mobility (MMC)

In June 2012, the federal government launched the Medium Cities Mobility Growth Acceleration Programme (PAC), with the aim of improving traffic in cities with populations between 250,000 and 700,000. The R$7 billion would be released to 75 medium-sized cities selected through a selection process. To do this, the municipalities had to draw up executive projects to carry out the urban mobility works and could submit up to two proposals (MINISTÉRIO DAS CIDADES, 2012).

On 8 January 2015, Joinville City Hall signed a R$105 million contract with Caixa Econômica Federal. The funds will be used for the largest investment in Joinville's history in projects to upgrade the urban road system, as part of PAC 2 MMC (the second phase of the programme). The municipality will be obliged to invest 5.2 million reais in return. Once the contracts have been signed, the next steps for the municipality will be to put the work out to tender and carry it out over the next 36 months (JOINVILLE MUNICIPALITY, 2015).

4.4.2. The proposal

The proposals for the construction of exclusive bus lanes drawn up by Joinville City Hall to improve urban mobility in the municipality are divided into two major projects, proposal 1 and proposal 2, where the first qualifies road axes in the centre, south and east of the city and the second qualifies the north and south axes. In Figure 5, proposal 1 is represented in red and proposal 2 in blue.

The full map of the programme can be found in the annexes to this paper (Annex 1).

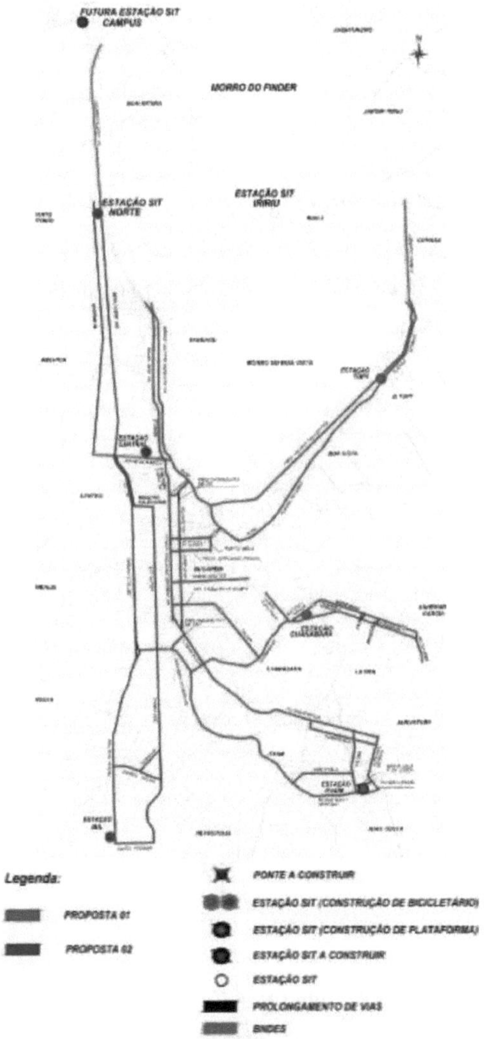

Figure 5 - **"Map of the PAC 2 DEC 2014 programme with anemones".**

Source: IPPUJ (2014b).

According to the proposals, 55 kilometres of roads on which urban public transport buses travel will be favoured by PAC 2 MMC. The projects emphasise the upgrading of the city's main north/south/east roads, which will receive concrete paving, resurfacing, pavements, cycle lanes, shelters and bike racks, integration stations and road signs (PREFEITURA MUNICIPAL DE JOINVILLE, 2015).

The proposal includes intermodal integration through the installation of 134 paracycles (bicycle racks without a cover), a new integration station near the University of the Joinville Region (Univille), with covered bicycle racks monitored by cameras at this and the other stations. The favoured roads will be given exclusive public transport lanes, cycle tracks and new passenger shelters. Binary systems involving Urussanga/Procópio Gomes and Hermann Leppes/Beira-rio streets will also be implemented (PREFEITURA MUNICIPAL DE JOINVILLE, 2015).

Mayor Udo Dõhler stressed that "We are doing serious work, planning does not allow for improvisation. As well as thinking about Joinville in 15 years' time, we are considering the growth we will have in 30 years' time." According to IPPUJ (2013), urban public transport accounted for 21.79% of journeys (data from the Origin-Destination Survey, 2010) and the goal is for this figure to reach 40% by 2030. The aim is therefore to prioritise urban public transport, making journeys faster and encouraging people to use it.

Given the importance of the proposal for the city, it was decided to analyse the project in order to identify its potential and the improvements needed to achieve efficiency in the public transport service. Section 5.2.3 specifies the focus of the analysis and the reasons for this choice.

4.4.3. The selected axis

In this study, we analysed the axis named East, referring to proposal 1, in red (Figure 5). For this study, we analysed the roads called Beira-rio and Procópio Gomes, which are linked to the eastern axis. These have individual executive projects which are referred to throughout the work as 'the projects'.

Based on the opinion[5] drawn up by IPPUJ (2013), Joinville's SIT carries 143,989 passengers a day. The routes on the proposed eastern axis carry 40% of the users of the entire integrated public transport system, which equates to 57,596 passengers a day.

In addition, according to IPPUJ (2013), a peak hour is equivalent to 10% of the day's total traffic. Five peaks were identified throughout the day (Table 3), resulting in seven hours and forty-five minutes of peak time, so it can be concluded that 77.5 per cent of passengers (44,637 passengers) travelling along the eastern axis are travelling at peak times.

Table 3 - Passengers transported per peak.

Order of peak	Timetable	Reason for	Quantity (%)	Total passengers	Total passengers

[5] IPPUJ - Fundação Instituto de Pesquisa e Planejamento para o Desenvolvimento Sustentável de Joinville (Org.). **Economic and socio-environmental justification for interventions on the eastern axis - proposal 01 - PAC medium mobility cities**. Joinville: City Hall, 2013.

occurrence		travelling		carried on the day (%)	carried on the day
1°	6:30 - 8:30	study trip (morning) going to work (6h) journey to work (8h)	3,50% 3,75% 33,75%	41%	18.301
2°	11:30 - 12:30	return study (morning) study trip (afternoon)	3,50% 1,50%	5%	2.232
3°	13:30 - 14:15	return to work (6-hour day)	3,75%	3,75%	1.674
4°	16:00 - 19:30	return to study (afternoon) study trip (evening) return to work (8am)	1,50% 7,50% 33,75%	42,75%	19.082
5°	22:15 - 22:45	return to study (evening)	7,50%	7,50%	3.348

Source: IPPUJ (2013).

Considering the importance of the eastern axis for Joinville's integrated public transport system, the projects for the roads belonging to it were analysed and it was observed that the only axes where there will be exclusive segregated corridors are those called Beira- rio and Procópio Gomes.

The Beira-rio axis covers the stretch known in the city as Avenida Beira-rio, which includes Avenida José Vieira and Avenida Doutor Albano Schulz, between Rua Nove de Março and Rua Padre Antônio Vieira, with an approximate length of 2,110 metres. Avenida Doutor Paulo Medeiros, which also benefited from the PAC2 MMC proposal, was added to the axis to give continuity to the corridor, as it connects the Beira-rio and Procópio Gomes axes with a length of approximately 460 metres, resulting in a total length of 2,570 metres. The Procópio Gomes axis, on the other hand, comprises only Rua Coronel Procópio Gomes with an approximate length of 1670 metres.

5.2.4 Population demands

In 2014, EMBARQ Brasil carried out the QualiÔnibus Programme Satisfaction Survey in Joinville. This was coordinated by the Joinville Sustainable Planning and Research Institute Foundation (IPPUJ), carried out by students from the Federal University of Santa Catarina (UFSC) Joinville campus and financed by FedEx Corporation.

The methodology consisted of 20 evaluation topics, including satisfaction (16), agreement

(2), problems (1) and evolution (1), which cover all aspects of the system's operation. The survey had a sample of 718 interviews, its universe was Joinville's public transport users and it had the help of 23 interviewers. It was carried out between 08 October and 07 November 2014. The overall aim was to understand the needs of public transport users in the city, as well as to unify surveys of this type carried out in more Brazilian municipalities and stimulate the creation of quality management systems in cities (WRI BRAZIL, 2017).

The results of the survey showed the dissatisfaction of Joinville's public transport users. The overall satisfaction score was 5.4 on a scale of 0 to 10. The highest scores for ease of payment, user service and user information were 6.7, 6.5 and 6.1, respectively. The lowest scores were for comfort at bus stops, cost and noise exposure, 3.4, 3.4 and 4.3 respectively (EMBARQ BRAZIL, 2014). This dissatisfaction with the SIT and the expectation generated by the promises and publicity of new projects are echoed in the local media.The dissatisfaction of SIT users is also evidenced by the decline in the number of users. Figure 6 shows the reduction in users over the last 30 years.

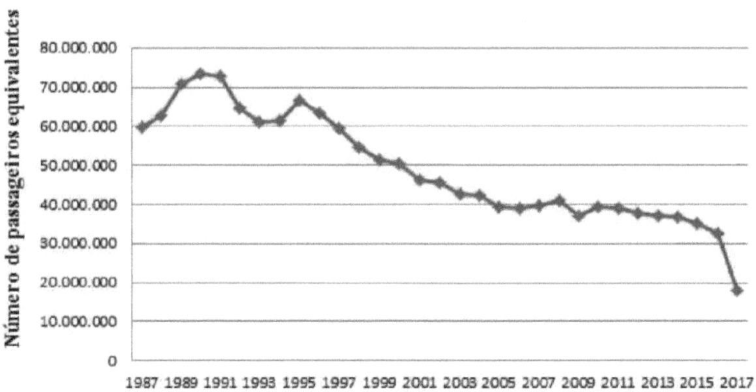

Figure 6 - Decline in public transport users over the last 30 years.

Source: Data provided by the Passebus company to SEPUD; graph authored by the company itself (2017).

It can be seen in Figure 6 that in 30 years there has been a reduction of approximately 40,000 users and that, considering the increase between 1987 and 1991, this number rises to around 50,000 users. The urgent need for investment in Joinville's SIT is evident from the sharp drop between 2016 and the last data collection carried out in 2017, where the number of users fell from approximately 32,000 to 18,000.

On 7 October 2013, the local media channel ANotícia published an article **entitled "Avenida Beira-rio de Joinville changes to have a BRT corridor" with the subtitle: "The** Bus Rapid Transit project, which starts next year, is part of a **R$104.8 million** investment by **PAC 2 and will upgrade 55 km of streets". This shows** the expectations generated by the proposal made by Joinville City Hall to the Federal Government.

Throughout this work, we have seen that the news item overestimates the proposal, as it only deals with the construction of exclusive lanes and not the implementation of a new system, a fact that can also be seen in its presentation in section 5.2.2. The funding was only signed in 2015, as described in section 5.2.1, despite the fact that the works were planned in the news item for 2014.

News outlets respond to public demand and sometimes generate overestimated expectations. For these reasons, the following section is necessary. It analyses the Beira-rio and Procópio Gomes axes, the only ones with proposed segregated lanes, located on an important axis of the city, as seen in section 5.2.2. By means of analyses, we seek to identify the potential of the proposal and the improvements and/or studies needed to achieve efficiency in its service and, in this way, meet the population's wishes.

4.5. COMPARISON OF THE PROJECTS FOR THE SELECTED ROADS WITH A BRT SYSTEM

Analysing the exclusive corridor projects for the Beira-rio and Procópio Gomes axes as a single project, it can be seen that they are similar to a BRT system due to their length and the presence of segregated lanes. Unlike the other corridors in the city, the segregated lanes are not shared by taxis, school vans, cars during conversion and emergency vehicles. Given the decline in the number of public transport users in Joinville and the population's demand for a higher quality and more efficient system, existing projects are being analysed with a view to a fast, comfortable and cost-effective transport system.

Given this similarity to a BRT system, when the individual projects of the axes are treated as a single project, they are referred to throughout the **work as 'the project'** characterising both axes under study.

4.5.1. Comparison in the light of the QBRT document

The PQBRT document, described in section 4.1, can be used as a tool for evaluating BRT systems, as it provides a quantitative assessment. With this in mind, the Joinville project was analysed using the scoring system proposed by the tool.

Initially, the essential elements that qualify a transport system as a basic BRT were highlighted. Remember that this is a basic qualification for the system to be assessed as gold, silver or bronze. And if it is not classified in any of the three categories, the basic BRT is still considered a BRT by the PQBRT document. Table 4 compares the items in the Joinville project with the essential elements proposed by the so-called PQBRT document and their respective scores. It's worth noting that the original tables can be found in section 4.1.

Table 4 - Essential elements, existence and punctuation.

Essential elements	**Existing**	**Score**	**Minimum required**	**Category minimum**	**Category maximum**
a) Segregated infrastructure with a priority of ticket*	yes	8	4	0	8
b) Alignment of bus lanes**	yes	8	4	0	8
c) Charging outside the bus	**no**	0	does not have	4	8
d) Treatment of intersections	yes	7	does not have	2	7
e) Level boarding	**no**	0	does not have	6	7
Total		23	**20**	12	38

Source: Author (2017).

a) Segregated infrastructure with priority of passage: analysing the first essential element in Table 4, it was found that the project has two types of separation with different scores. In totalling the final score, these points per infrastructure will be multiplied by the percentage of the corridor that has each type of separation, as can be seen in Table 1 of section 4.1. This score is calculated at the end of analysing the elements. Eight points were awarded due to the predominant length of the Beira-rio axis at 2570 metres, which has physical separation. The length of the Procópio Gomes axis is 1670 metres with a segregated lane, separated by a line painted on the pavement on the left-hand side. This stretch is not isolated by physical separations and scored four points in the infrastructure element (IPPUJ, 2013). Figures 7 and 8 show the type of separation, the difference in alignment and the treatment of intersections, also essential elements that will be discussed below.

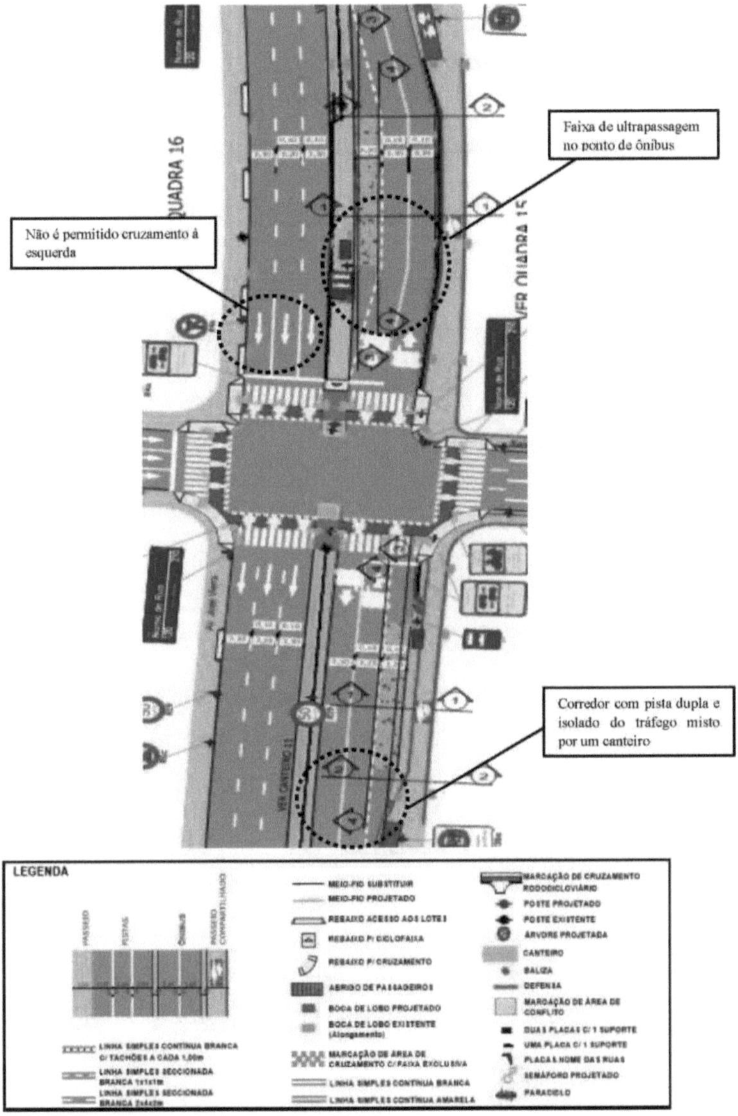

Figure 7 - Section of the Beira-rio signalling project, intersection of Av. José Vieira and Itaiópolis.
Source: Adapted from IPPUJ (2015b).

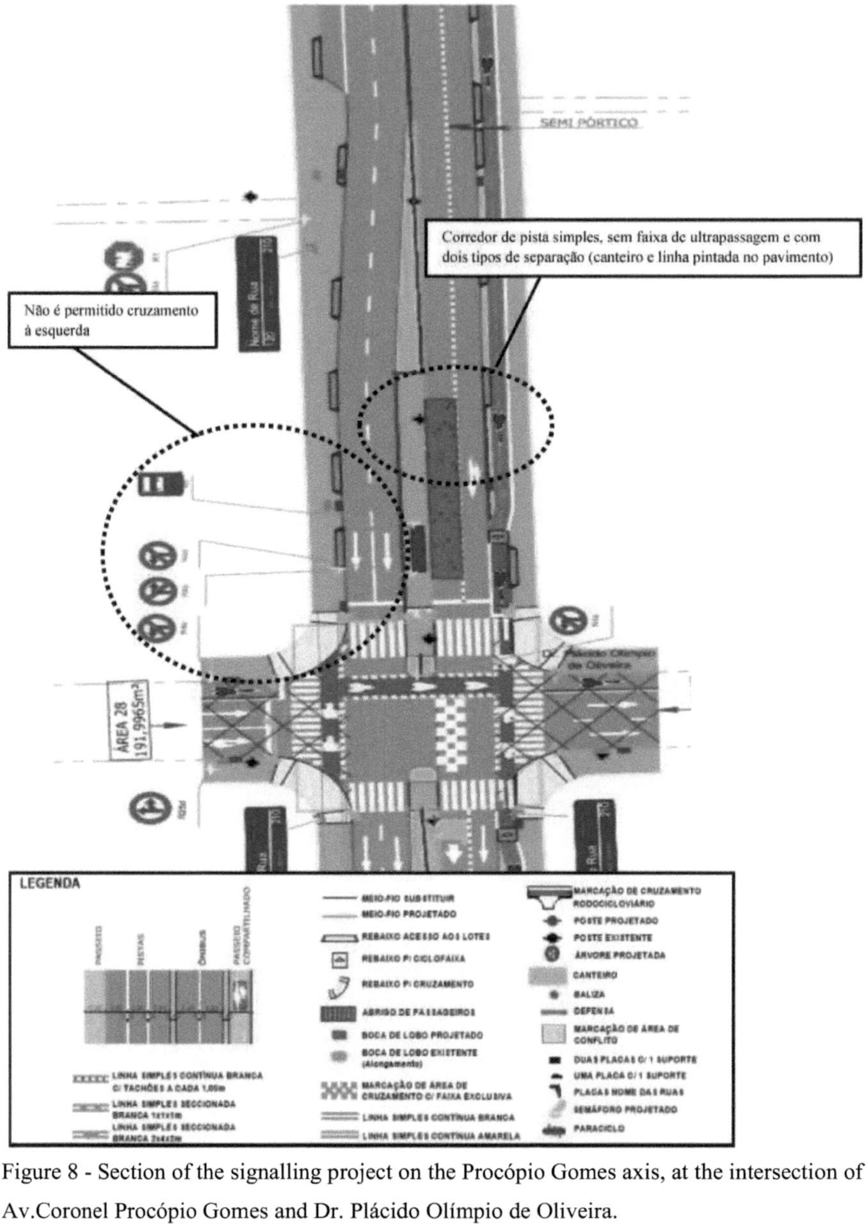

Figure 8 - Section of the signalling project on the Procópio Gomes axis, at the intersection of Av.Coronel Procópio Gomes and Dr. Plácido Olímpio de Oliveira.

Source: Adapted from IPPUJ (2015c).

b) Alignment of bus lanes: analysing the second essential element in Table 4, it can be seen that the axes under study again differ in terms of infrastructure. On the Beira-rio axis, as the name suggests, the avenues border a river, characterising the alignment as level 1 with bus lanes running adjacent to river banks, where there are few intersections that could cause conflict, scoring eight points, as shown in Table 2 in section 4.1. However, the Procópio Gomes axis is not on the river bank and has mixed vehicle lanes on its right and left sides, defining it as level 2. It also has a bus lane aligned centrally on the one-way street, so it scored five points, also following Table 2. The same scoring criterion used for the previous element was adopted in relation to the length of the axes, so the value in Table 4 refers to the Beira-rio axis.

c) Charging outside the bus: in relation to the third essential element in Table 4, the BRT Quality Standard document defines three types of charging that facilitate boarding and alighting, which reduce journey times: access control, proof of payment and on-board ticket validation (all doors). The shelters on the axes studied will not be characterised as stations; access control will continue to take place only at the terminals and integration stations already established in Joinville. Ticket validation will continue to take place on board, but, unlike the PQBRT document, only at the first door of the vehicle. As such, it does not score on this element and lacks an essential element for the corridor to qualify as BRT.

d) Treatment of intersections: the fourth line in Table 4 shows that this essential element has the maximum score, as conversions will be prohibited on bus lanes. The images in Figures 7 and 8 show cut-outs of the signalling project, displaying the signs and horizontal markings that will guide drivers. At intersections where the bus needs to leave the corridor, a new traffic light priority project is planned, but even without this, the project receives seven points in the treatment of intersections.

e) Level boarding: this is another essential element not present in the Joinville project, as the horizontal and vertical spans will not meet the defined standard. The use of **low-entry** buses reduces the vertical span and facilitates boarding, which is called BRS. Unlike the BRT, the BRS is used in shelters that do not have a boarding platform, making it easier for users to board due to the lack of stairs inside.

Other elements are assessed in the PQBRT document and are present in the exclusive corridors proposed on the axes studied, but will not be analysed separately as they are discussed in section 5.3.2. These deal with service planning, infrastructure, communication and access and integration. Table 5 highlights the existence of these elements and their scores, as well as the essential elements assessed previously for better visualisation and calculation of the final score.

Table 5 - Total score.

	Element	Existing	Score	Description	Weighting
	Segregated infrastructure with priority passage	yes	6,44	Segregated and physically separated lanes / Segregated lanes separated by a line painted on the pavement.	8 * 61%/ 4 * 39%
Basic BRT	Alignment of bus lanes	yes	6,83	Bus lanes that run adjacent to riverbanks where there are few intersections that could cause conflict	8 * 61%/ 5 * 39%

				(level 1) / Bus lanes divided into pairs of one-way lanes on separate streets, with each bus lane aligned centrally on the road.	
	Fare collection outside the bus	no	0	None.	-
	Treatment of intersections	yes	6,44	Prohibited conversions in bus lanes.	7 * 92%
	Level boarding	no	0	It does not.	-
Service planning	Multiple lines	yes	4	There are two or more lines in the corridor that serve at least two stations.	-
	Limited and local express services	yes	2	At least one local service option and one limited or express service option.	-
	Control centre	no	0	No control centre or the centre has limited functions.	-
	Location in the top ten corridors	yes	2	The corridor is one of the ten corridors with the highest demand.	-
	Demand profile	yes	3	The corridor includes the segment with the highest demand and has a Level Corridor configuration 1.	-
	Timetable Operation	yes	2	It has evening and weekend service.	-
Infrastructure	Overtaking lanes at stations (shelters on the Beria Rio axis)	yes	3	Segregated overtaking lanes	-
	Overtaking lanes at stations (shelters on the Procópio Gomes axis)	yes	1	Overtaking in mixed traffic lanes safely.	-
	Minimising bus emissions	no	0	Levels below the expected standards.	-
	Stations away from intersections	no	0	Distances not foreseen in the project.	-
	Central stations	no	0	There are no central stations.	-
	Floor quality	no	0	The useful life of the pavement is less than 30 years.	-
Stations	Distances between stations	yes	2	Stations spaced on average between 300 and 800 metres apart.	-
	Safe and comfortable stations	yes	2	Stations have 2 elements (width and weather protection).	2* 100%
	Number of doors on the bus	no	0	Board only through the front door.	-
	Shoulder bays and stop sub-points	no	0	There are no sub-parking points or shoulder bays (a bay with space for just one vehicle).	-
	Sliding doors at stations	no	0	No sliding doors.	-
Communications	Brand consolidation	no	0	There's no corridor marking.	-
	Passenger information	no	0	There is only static information on the terminals.	-
Access and Integration	Universal Access	yes	2	There is physical accessibility.	-
	Integration with other modes of public transport	no	0	No integration.	-
	Road safety and access	yes	4	Good and safe pedestrian access at all stations and many improvements along the corridor.	-
	secure bicycle parking	no	0	Little or no parking for bicycles.	-
	Cycle infrastructure	yes	1	Cycle infrastructure does not accompany the entire corridor.	-
	Integration with bike-sharing systems	no	0	There is no bike-sharing system.	-
Total			47,71		

Source: Author (2017).

The classifications for a BRT system proposed by the PQBRT document stand out: Basic BRT (mandatory presence of the five items said to be essential for the system and obtaining at least 20 points in these), Gold Standard BRT (85 points or more), Silver Standard BRT (70 - 84.9 points) and Bronze Standard BRT (55 - 69.9 points). Even exceeding the minimum score (Table 4), the corridors studied are not characterised as a Basic BRT system due to the lack of off-board charging and level boarding.

However, with a final score of 47.71 points (Table 5), the result was close to the minimum for a BRT system classified as Bronze (55 points), which reaffirms the potential of the corridors studied. From the low-scoring or non-existent elements, we can highlight what needs to be improved in the project in order to make it more efficient.

4.5.2. Comparison in the light of the MBRT document

The projects proposed by public bodies seek to improve urban mobility and comparing them with the MBRT document makes it possible to identify improvements or studies needed to make them successful. Exclusive lanes prove to be efficient when installed in the right location to meet an expected demand, and have increased potential if fully segregated.

With a view to efficiency, the proposal under study is compared with a higher-capacity, high-quality system by means of the MBRT document, which provides a conceptual assessment. For this section, the analyses are sequenced according to the order established in Chapter II of the MBRT document - Operational Design: 5.3.2.1 Network and line design; 5.3.2.2 System capacity and speed; 5.3.2.3 Intersections and traffic light control; 5.3.2.4 User service.

4.5.2.1. Network and line design

To better organise this section, it has been divided into the following items: a) Characterisation of access to the corridors, b) Identification and discussion of the service provided, c) Description of the lines and transfers and d) Temporal integration and line drawings.

a) Characterisation of corridor access:

The corridors implemented in Joinville are open systems, through which public transport vehicles controlled by the operators, taxis with passengers, school vans, emergency vehicles and private vehicles for conversions pass. However, the routes selected for study will be closed systems once they are implemented, limiting access only to the operators Transporte e Turismo Santo Antonio S.A (Transtusa) and Gidion Transporte e Turismo S.A.

According to the MBRT document, in general, closed systems allow for greater fluidity, as there will be less interference, leading to an increase in average speed and a reduction in users' journey times. This will also allow the vehicles selected to travel on the corridors to be specific, in which case BRS buses, the system defined in section 2.1.2, should be used to ensure greater accessibility to the system.

b) Identification and discussion of the service provided

The operational logic of Joinville's public urban user transport service, as an integrated system, is based on a trunk feeder service. The MBRT document **states that "trunk feeder** services use smaller vehicles in low-density areas and larger vehicles along corridors in high-density areas. The **smaller** vehicles **therefore 'feed' the larger 'trunk' corridors" (MINISTÉRIO DAS** CIDADES, 2008, pg.236).

The operational structure is organised into 10 integration stations (terminals) connected by 21 trunk lines (14 stop lines and seven direct lines), 17 interstation lines, four neighbourhood lines and 184 feeder and special lines. The latter are conditioned by the main road system and the existing road traffic system. As a result, the network offers 247,711 journeys per month, with 1,855,076 kilometres on the road, and a total fleet of 351 buses (IPPUJ, 2014a).

According to the MBRT document, Ministry of Cities (2008), there is no correct decision between trunk or direct services. Each option becomes efficient according to the characteristics of the environment where it will be deployed. The MBRT document mentions some conditions in which the service is likely to be efficient.

With regard to the first condition, it should be noted that the main corridors cater for high demand, which has already been verified on the axes under study. The second important condition is the difference in population density between different areas of the city. Figure 9 shows this situation, as despite greater densification in the centre of the map, the densities are very different.

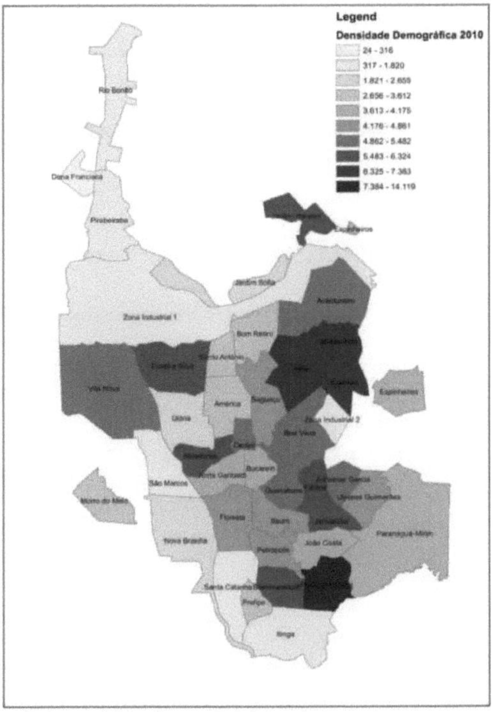

Figura 9 - Population density of the city of Joinville year 2010.

Source: IPPUJ (2010a).

The last condition pointed out by the MBRT document relates to the distances between the city centre and feeder areas, around 10 kilometres. Figure 10 shows Joinville's network of public transport lines, differentiating them by colour: green for feeder lines, red for trunks and blue for interchanges. The boundary of the municipal area is highlighted by the black line, highlighting the long distances covered by the feeder lines.

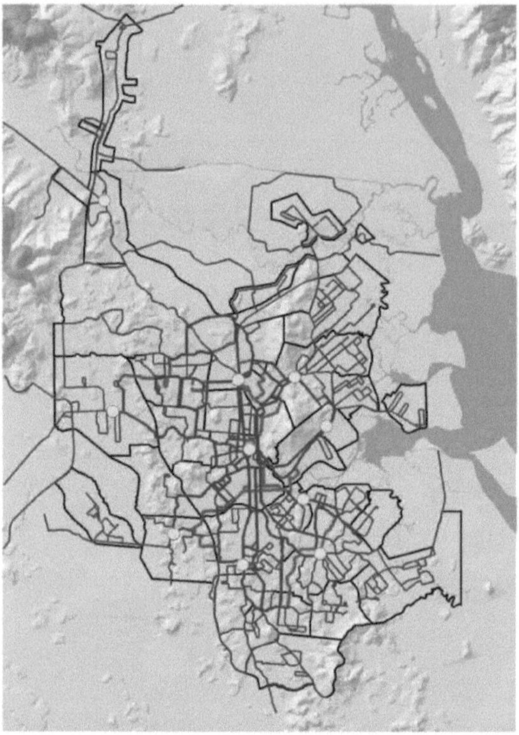

Figura 10 - Map of Joinville's Integrated Transport System.

Source: SEPUD (2017b).

Thus, the three conditions proposed by the MBRT document for the probable success of the trunk feeder service are met in the Joinville SIT. It is worth emphasising that the axes under study have the potential for an efficient service, but the system must work as a whole. The lack of exclusive corridors connecting the terminals is detrimental to the quality of the service, as it increases the user's journey time by using roads that are more influenced by private vehicles.

c) Description of lines and transfers

By analysing the current bus stops, located on the streets of the Beira-rio and Procópio Gomes axes, and comparing them to the proposed shelters (Figures 11 and 12) and the existing projects in the PAC 2 MMC proposal, it is possible to define which lines will be served by the corridor during its implementation. Figures 11 and 12 show the location of the shelters and their service radius. The full proposal can be found in the appendices to this paper (Appendix 2).

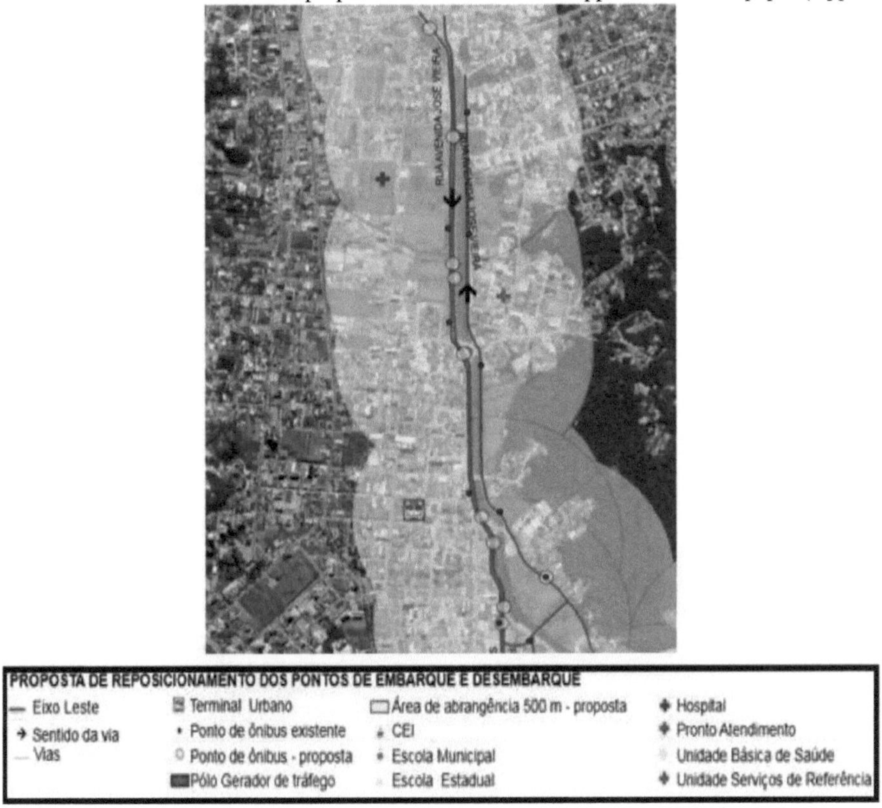

Figura 11 - Proposed shelters where the Beira-rio axis will be located.

Source: Extract from "Figure 01 - Proposed repositioning of boarding and alighting points"; IPPUJ (2014d).
disembarkation points"; IPPUJ (2014d).

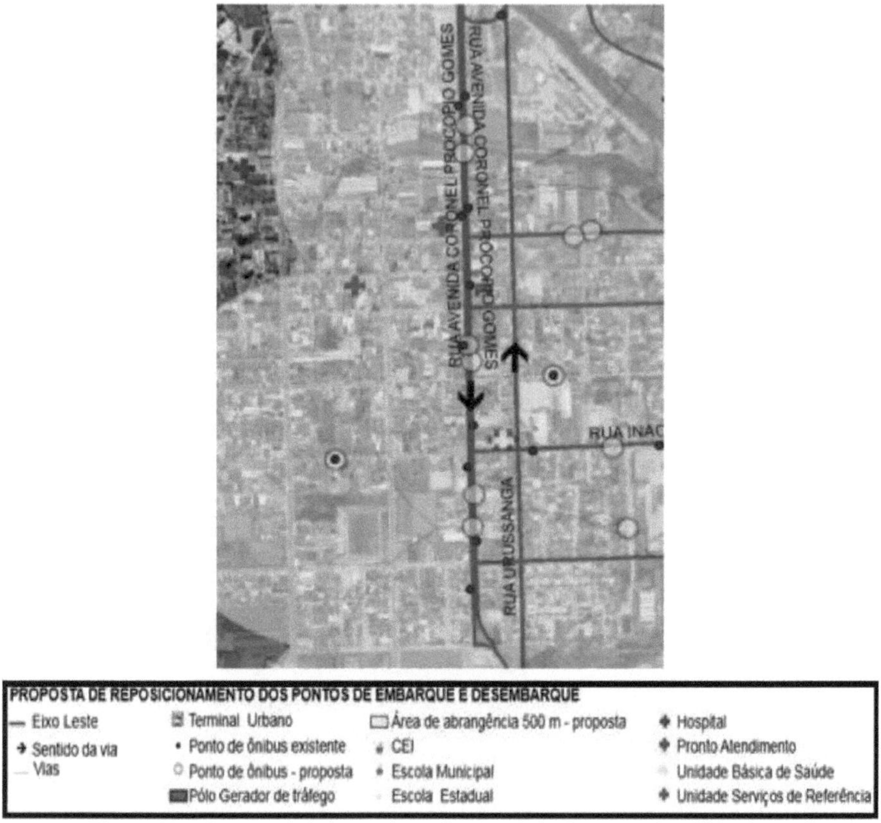

Figura 12 - Proposed shelters where the Procópio Gomes axis will be located.

Source: Extract from "Figure 01 - Proposed repositioning of boarding and alighting points"; IPPUJ (2014d).

Some items can be discussed on the basis of Figures 11 and 12, such as: traffic generating centres, the range of services provided by the shelters, among other elements. In this case, the focus of the case study is an analysis of the positioning of the bus stops and the lines they will serve. The distances and proximity between the new bus stops do not match the alignment of the roads in the project.

The Procópio Gomes axis will have one-way streets for both the corridor and the mixed carriageway, which is why it cannot be said that the proximity of the points means that they are positioned in different directions. However the Beira-rio axis will be one-way on the dual carriageway and two-way on the bus lane, which would justify the close points, i.e. one for each direction.

The proposal for shelters dates from 2014 and no updated version was found in the secondary data search. The project, on the other hand, is from 2015 and the comparison made between the proposal and the alignment of the roads shows a lack of interaction and involvement between the professionals at the different stages of the project.

In the survey of the lines served, the points located on the current route were analysed using the Onibus.info and Google Maps tools. Taking into account the changes in direction, the lines that would continue to run along the axes were included in Tables 6 and 7. These tables show the lines surveyed, as well as the axis where they are located, their classification (trunk or feeder) and the direction of travel.

Table 6 - Surveyed lines: Procópio Gomes axis.

Axis	Lines	Classification	Sense
	304		Itaum / Centre
	305		Itaum / Campus
	306		Itaum/ Campus via Guanabara
	307		Itaum / Pirabeiraba
Procópio	600		Guanabara/Centre
Gomes	1402		Procópio Gomes Circular
	1410		Evening Circular Estevão de Matos
	1704		Itaum / Dohler
	1712		CEDUP / Centre
	1720		Celso Ramos College / South
	Troncal		Feeder

Source: Author (2017).

Table 7 - Surveyed lines: Beira-rio axis.

Axis	Lines	Classification	Sense
	40		Tupy/ North via Centre
	43		Tupy/ Centre
	135		North/ Centre via Dona Francisca
	244		Benjamin Constant/ Centre
	245		Orestes Guimarães Circular
	304		Itaum / Centre
	306		Itaum/ Campus via Guanabara
	307		Itaum / Pirabeiraba
	601		Guanabara/Centre
	800		Iririu/ Centre
Riverside	802		Iririu/ Centre via Castro Alves
	806		Iririu/ Semi-direct centre
	1402		Procópio Gomes Circular
	1410		Evening Circular Estevão de Matos
	1704		Itaum / Dohler

	1712		CEDUP / Centre
	1721		Celso Ramos College / Centre
	1734		Itaum / Lepper via Guanabara
	1742		Senai/ Centre
	2010		Circular Centre
	2015		Viewpoint
	Troncal	Feeder	

Source: Author (2017).

In tables 6 and 7, the lines highlighted in red are classified as trunk and direct, and the lines highlighted in green as feeder lines. According to the IPPUJ[6] (2014a), trunk and direct lines serve the main roads connecting the central areas. The former have a stop every 400 metres on average and the direct lines have a maximum of one or two intermediate points between the centre and the neighbourhood stations. The so-called direct lines that stop at intermediate points are actually **called semi-direct lines. An example is** *THE* **"Iririu/Centro semi-direct" line, which stops at two** points located on Avenida José Vieira (Beira-rio axis). This is the only one with

semi-direct classification, the other red lines are trunk lines, or stops, as defined in IPPUJ (2014a).

There are a total of 21 trunk lines with 77,624 journeys per month. Of these, 12 lines, or 57 per cent, will run along the Beira-rio and Procópio Gomes routes. However, there are 184 feeder lines and 13 of these will pass through the axes (only 7%).

The feeder lines, with their capillarity, have circular and radial routes to the integration stations. They cover the greatest total kilometres of the system, offering 142,849 journeys per month. Because they are close to residential areas and do not have priority for circulation, the feeder lines need improvements to the infrastructure of the roads used and greater clarity in the road hierarchy. The latter is necessary so that the speed established on the road fulfils its function (local, collector or expressway).

Despite the close number of trunk and feeder lines served by the axes, 12 and 13 respectively, when looking at the ratio to the existing total (12/21 trunks and 13/184 feeders) it can be concluded that the corridors will meet considerable demand, characterising a high-capacity system.

However, it is not ideal for feeder lines to also use the corridors, as buses stop more frequently (as we have seen, they need closer stops) and use smaller vehicles. The more frequent stops can interfere with the movement of the trunk bus, when there is no overtaking lane as on the Procópio Gomes axis.

The problem would be solved by limiting feeder lines to residential areas or more distant areas, where the user would need to transfer to the trunk line. However, this approach may displease

6 IPPUJ - Fundação Instituto de Pesquisa e Planejamento para o Desenvolvimento Sustentável de Joinville (Org.). **Joinville's mobility and accessibility plan (PlanMOB)**. Joinville: City Hall, 2014a, 107 f.

users, as they tend not to like transfers. A transfer requires the user to travel, which can be a nuisance for people with walking difficulties or toddlers, for example, as well as infrastructure costs for transfer stations. As such, transfers are only welcome if they are easy and comfortable, without generating long journeys, increased costs or unpleasant experiences for the user. The transfer is provided for in the Joinville SIT and is detailed below in item d.

The mix of lines using the exclusive corridor is due to its location, as there are no transfer terminals at the beginning or end of the corridor. In this way, when accessing the terminals, the bus will use preferential (non-exclusive) lanes, reducing the efficiency of the service. Returning to the proposals made to the PAC 2 MMC, described in section 5.2, we emphasise the importance of implementing corridors and ensuring that they are exclusive so that the system is truly efficient. Figure 13 is a cross-section of the presentation of the Eastern Axis proposed for PAC 2 MMC. The routes highlighted in green are those with an exclusive corridor, while the other routes highlighted in red will be preferential.

Figure 13 - Cut-out of the figure **"EAST AXIS".**

Source: IPPUJ (2014c).

Figure 13 shows that the corridor highlighted will not connect the integration terminals. This highlights the need for exclusive lanes on the other routes shown in red. However, preferential corridors are planned.

Despite the decrease in service quality due to the proposed corridors shown in red, the analysis of the one shown in green continues. The efficiency of the corridor under study is necessary to improve the public transport service in Joinville.

d) Time integration and line drawings

In order for a network of lines to be efficient, it is necessary to: minimise transfers by reducing journey times, offer local and express services so that users can choose the line that best satisfies their reason for travelling and not make unnecessary journeys, and shorten some lines within the bus corridor by concentrating them in the sections with the highest demand (MINISTÉRIO DAS CIDADES, 2008).

Integration in the Joinville system takes place mainly at the terminals. At the bus stops, there is temporal integration: the user doesn't pay for the second ticket within an hour of using the first one if they have the so-called Ideal Card (non-returnable) and if the lines integrate with each other in a single direction. For example, if the user remembers on the bus that he has forgotten his mobile phone at home and decides to go back for it, he will pay the second fare. The ticket is free if the user chooses another line that runs in the same direction. An example of this would be leaving a feeder line and getting on a trunk line that runs in the same direction in order to use a system that allows for a shorter journey time.

However, there is no map of the integrated system for users. The existing map is Figure 10, which shows the integration of the network, but there is no identification of the lines or shelters. Users need to look up information on line identification and direction on mobile apps.

The need to make transfers and the lack of information for users jeopardise the journey and thus user satisfaction. As the system has a trunk feeder service, transfers are inevitable, which is why the design of networks and lines must take care not only to meet demand, but also to provide for transfers outside terminals and better inform users. However, the line project is not included in the proposal submitted to PAC 2 MMC.

As for the express and local services discussed in item b, it is known that they are present in the system. The shortening of lines is not addressed in the proposed project.

5.3.2.2 System capacity and speed

The efficiency of an integrated transport system revolves around three elements: demand, capacity and speed. In section 5.3.1, we discussed the infrastructure needed to meet demand, high-capacity routes and achieve a shorter journey time. Section 5.3.2.1 has the same objective, but with an operational approach to the lines and the network. This section compares the calculations carried out for the project under study and the calculations proposed by the MBRT document to quantify the system's capacity and speed.

When analysing the three volumes of the proposed PAC 2 MMC project, which are made up of studies, execution projects and specifications, it was found that they only deal with infrastructure issues. We therefore used the calculations proposed in the 2013 Call for Tenders[7] , hereinafter referred to as the Call for Tenders.

The tender defined the formulas for vehicle capacity, the renewal rate, the number of departures and the calculation of the minimum fleet required. When calculating the total capacity of the vehicle, the parameters used are the number of seats in the vehicle, the usable area available and the number of standing passengers per square metre, which can be seen in Table 8. The renewal index is the ratio between the total number of passengers carried on the journey (outward and return) and the maximum capacity of the vehicle reached on the journey. The number of departures is important for determining the frequency of a line, characterising the number of vehicles needed to meet demand and this depends on demand in the time slot, the renewal rate and the total capacity of the vehicle.

The MBRT document, on the other hand, defined more complete and specific formulas for the corridors. According to it, the infrastructure, the choice of vehicle size and the number of lines depend on the demand they will meet and characterise the capacity and speed of the corridor. To this end, he proposes formulas for corridor capacity in general, corridor capacity at a station, capacity based on vehicle length, the impact of vehicle length on dead time, determining the required vehicle capacity, operational fleet size for a corridor, saturation level in a stop bay, optimum number of lines, stopping time within a train system and optimum distance between stations.

Table 8 shows that both documents deal with vehicle capacity, but the MBRT document proposes two different formulae. The first formula calculates capacity according to the length of the bus, taking 10 people per metre as the length of the vehicle and 3 metres as space for the driver. The second formula is a relationship between corridor capacity, occupancy factor, service frequency (vehicles/hour) and number of stop bays at stations. The latter is more complete and takes operational factors into account, but should be used if the stop bays are not saturated.

Table 8 also shows the proposed formulas for the size of the vehicle fleet. As for the number of vehicles needed for operation, the public notice defined the maximum fleet as being determined by the number of departures needed in the time slot with the highest load (hour/peak). And the MBRT document related cycle time, vehicle capacity and demand along the corridor. In this calculation, the manual and the public notice are similar in that they take into account vehicle dimensions, passenger movement and cycle time.

Table 8 - Capacity and fleet formulas side by side.

Public Notice	MBRT

[7] SEPUD - Joinville Urban Planning and Sustainable Development Secretariat. **Call for Tenders n°##/ of 01 September 2013:** Annex II - Basic Project. Document available for consultation only. Joinville, City Hall, 2017b.

Capacity Total *Vehicle*	$CT = NA + (AxP)$	$NP_{bus} = 10x(C_{bus} - 3)$	$NP_{bus} = \frac{Cap_{corredor}}{(F_{ocup}xFreqxN_{par})}$
Fleet required	$F = \frac{TC}{ip}$	$N_{bus} = \frac{DxT_{ciclo}}{NP_{bus}}$	

Source: Author (2017).

While the public notice is only concerned with vehicle capacity, the MBRT document considers the capacity of the corridor in general and at a station. The capacity of the corridor is important in order to size it correctly, avoiding excessive operating costs and unnecessary investments.

The calculations of the impact of vehicle length on dead time, saturation level in a stop bay, optimum number of lines, stopping time within a train system and optimum distance between stations proposed by the manual are specific to corridors. Dead time refers to the part of the bus's stopping time due to the longer time needed to pull onto and off the platform. The saturation level of the stop refers to the time the stop is occupied and is calculated to avoid congestion at the stop. Stopping time and optimum distance are also factors that increase efficiency.

However, these calculations were not carried out for the project (Beira-rio axis and Procópio Gomes axis) under study, which is why they are so important. The corridor capacity calculation itself, which is important for dimensioning the corridor, undergoes changes throughout the MBRT document. The initial formulae are proposed with basic elements such as service frequency, vehicle capacity and number of stops, but in the course of this study it was seen that many infrastructure factors affect the capacity of the corridor, such as the type of structure for boarding and alighting, the number of doors on the vehicle, charging outside the vehicle, in short, several other parameters can be related to each other to calculate capacity more accurately. Whether to include them is up to the planner. The MBRT document gave several examples of calculations in existing BRT systems, which are a reference for future projects.

5.3.2.3. Intersections and traffic light control

This section identifies the measures taken in the project under study with regard to intersections. The MBRT document points to intersections as critical points in a BRT system and the same concern is needed in exclusive corridors, because in both cases it is necessary to minimise the wait for public transport users, improve the safety and convenience of access to bus stations for pedestrians and minimise waits for mixed traffic.

It can be seen that exclusive lanes are generally implemented where traffic congestion is a problem, or where congestion is likely to occur in the near future. Otherwise there would be no benefit in building a segregated bus lane. For the corridor to be politically successful, it must not only improve public transport, but also help the flow of mixed traffic (MINISTÉRIO DAS CIDADES, 2008).

In the projects for the Beira-rio and Procópio Gomes axes in question, there was no specific study for the treatment of intersections (identification of bottlenecks or traffic light control). It is only through the signalling projects for the axes that it is possible to verify the existence of prohibited turns for mixed traffic, to the right or left depending on the alignment of the road, as identified in section 5.3.1 (where the treatment of intersections was assessed, Tables 4 and 5).

With regard to traffic light control, SEPUD was asked about the possibility of changing the traffic light timings on the corridors. SEPUD informed us that a study is planned, but it is not expected to start. As a result, there will be no traffic light priority for buses, and traffic lights could become bottlenecks in the corridor.

In Joinville there are four types of traffic lights: Central, Isolated, GSM and Flashing. In the Central traffic light, the controller is connected to a computer, which allows traffic light times to be changed remotely from a control centre. The Isolated traffic light is one that has no relationship with the traffic lights around it, i.e. it operates independently (CUCCI NETO, 2015). The GSM (Global System of Mobile Communications) traffic light stands out for its wireless data transmission. The flashing traffic light is when the yellow light remains flashing, i.e. the junction is not interrupted, it just has to be carried out carefully.

Figure 14 shows the type of traffic light on the axes studied and shows that the central traffic light predominates in the region. As such, traffic light times can be altered at the control centre. The complete map showing the location of traffic lights in Joinville can be found in the annexes (Appendix 3). It shows the predominance of the central traffic light, which is important for the city because it facilitates traffic light control and enables synchronisation between the different traffic lights. There are also GSM traffic lights located sparsely throughout the city, which shows an interest in this type of technology. A worrying fact is the use of isolated traffic lights, which are detrimental to the planning of the road system.

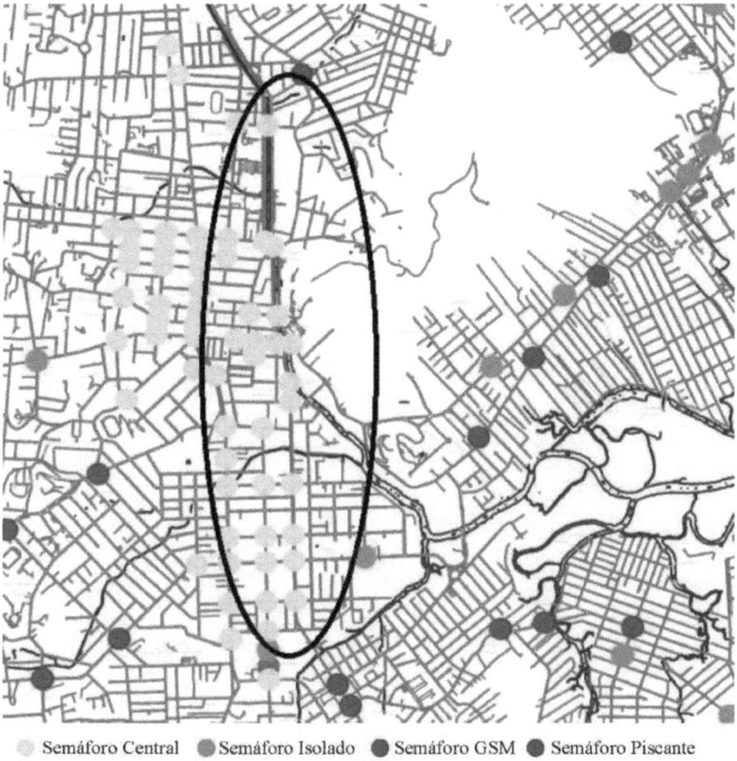

Figure 14 - Traffic lights on the axes studied.
Source: IPPUJ (2010b).

Traffic light control is not provided for in the project in question, but it is worth highlighting the discussions in the manual. The MBRT document points out the importance of evaluating the intersection to identify bottlenecks, proposes calculations for waiting at traffic lights, provides different cases and specifications for defining traffic light phases at conversions, emphasises the importance of thinking about cyclists and pedestrians, presents drawings for vehicle conversions, discusses the conversion lane dedicated to the bus to keep it in controlled spaces at all times, discusses the location of stops at intersections, calculates the levels of interference of the station at the intersection, discusses roundabouts and presents different types of traffic light priority (passive and active).

The project studied uses signalling rules defined by the Brazilian Traffic Signalling Manual (2007) proposed by CONTRAN (National Traffic Council). This prioritised conversions for public transport and was concerned with pedestrians and cyclists. If the signalling project is followed correctly, pedestrians and cyclists will be safe around the shelters, and there will also be integration between public transport and cycling. Figure 15 shows the surroundings of a shelter at an intersection located on the Beira-rio axis, highlighting the accessibility of public transport on the corridors and the signs for turning.

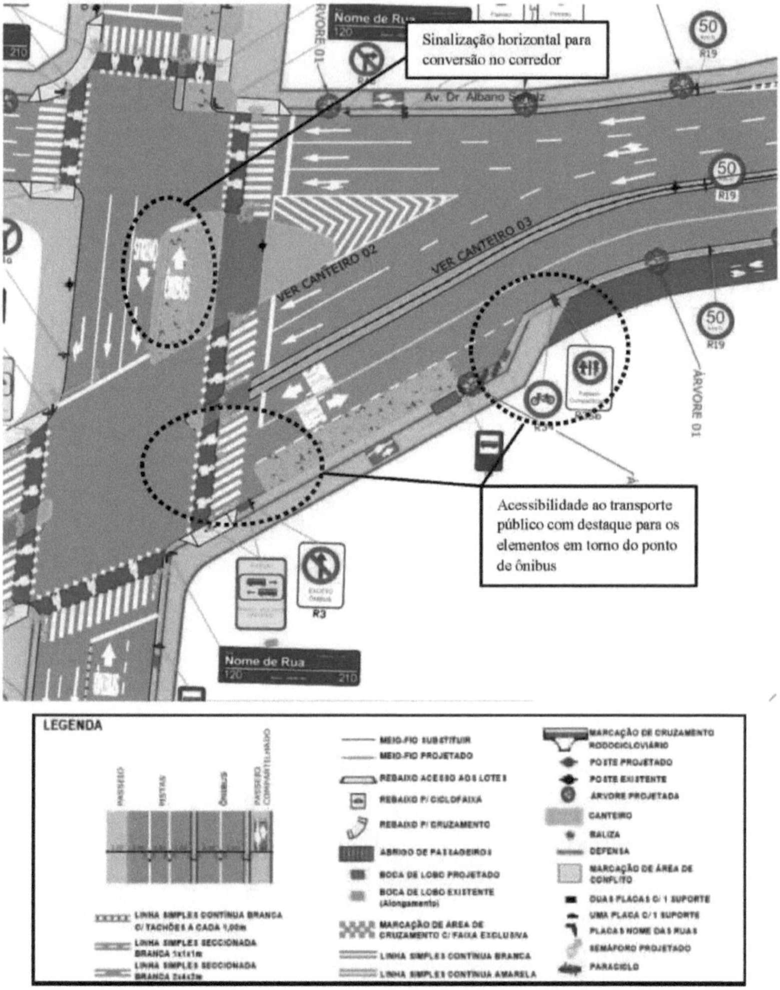

Figure 15 - Highlight of an intersection on the Beira-rio axis.

Source: IPPUJ (2015b).

The signalling projects addressed the following aspects pointed out by the manual: access for pedestrians and cyclists, designs for turning vehicles and the turning lane dedicated to buses. There was no specific study of intersections, but the safety of pedestrians, cyclists, users and drivers was ensured by the Brazilian Traffic Signalling Manual. What stands out is the lack of study into traffic light control, the location of bus stops (also discussed in section 5.3.1 where existing and proposed stops were compared) and interference with stations.

5.3.2.4. User service

This section elucidates the important points for providing a quality service to users according to the MBRT document and discusses the service provided in Joinville. The corridor project under study does not address the service that will be offered to users, so it is assumed that it will remain the same.

The MBRT document discusses the following elements:

a) User information by means of system maps, signage (in and around stations, as well as inside the vehicle), commercial advertisements to generate income for the system, public service messages, visual and voice information systems, interaction with users (user service centre, contact by e-mail and information sites, satisfaction survey);
b) Professionalism of the system, which also takes cleanliness into account;
c) Security;
d) Comfortable facilities;
e) Services division.

In order to qualify the service provided in Joinville, the satisfaction survey carried out in 2014, QualiÔnibus, was used to identify the quality of the service provided by the operators. Figure 16 shows the scores given by users to the Joinville SIT.

Figure 16 - Satisfaction survey.

Source: EMBARQ BRASIL (2014).

Despite the reasonable score for user information (6.3), obtained in the satisfaction survey seen in Figure 16, the SIT currently does not provide information about the lines or timetables at the bus stops, there is no real-time information and the information at the integration terminals is the minimum necessary, such as the name of the line and the scheduled departure time. The lowest scores are for the comfort of the bus stops and the cost, but it is worth noting that the system obtained an average score of 5.4 and the highest score was 6.7 (in the range of 0 to 10). This shows that users are dissatisfied with public transport in Joinville.

CHAPTER 6

PERCEPTION OF PROFESSIONALS IN THE FIELD

This section seeks to verify the perception of urban public transport professionals in relation to the analysis of the study by means of an interview. The script drawn up for the interview can be found at the end of this work as an Appendix. The problems observed during the case study were addressed in order to reinforce points and clarify the participation of SIT's operational planning sector in the preparation of public transport projects developed by Joinville City Hall.

According to Newman et al. (2011), during data interpretation it is necessary for the researcher to verify any inferences made before drawing a final conclusion. The authors also cite the Qualitative Legitimation Model proposed by Onwuegbuzie (2000), which attempts to integrate different types of validation identified by qualitative researchers and describes 24 methods for assessing the true value of interpretations in qualitative research. Of these, this section uses Peer Debriefing, in which the researcher contacts other professionals to get their perspectives on the researcher's perception of what has been experienced or observed.

The interviewees were chosen because of the area in which they work. They are employees of Passebus and work directly with the operational planning of Joinville's public transport system. Of the four participants, one is a manager in the administration and finance sector, and the other three work in operational planning.

According to the **company's** official website[8] **(2017), "Passebus** is a private company whose shareholders are Transtusa and Gidion. It was officially created on 2 January 2001 with the aim of coordinating the implementation of the ticketing system that would take place in Joinville in February of the same year and managing the **commercial** area." **In addition to the functions required for ticketing (sales, control, modernisation and**
special users), the company manages the database, the tele-service for users and the Citizenship Stations.

Given the importance of verification and the need to contact professionals in the field of study, section 6.2 characterises the interviewees in order to highlight the selection made, and section 6.3 presents the results of the interview.

6.2. CHARACTERISATION OF THE PROFESSIONALS

Table 9 shows the name, position, sector, length of time working in the area and training of the professionals interviewed in order to verify the analysis. The interviewees authorised the

[8] PASSEBUS (Joinville). **Institutional.** Available at: <http://www.passebus.com.br/>. Accessed on: 5 Nov. 2017.

disclosure of their data.

Table 9 - Characterisation of the interviewees.

Name	Position	Sector	Time working in the area	Training
Alamir Schelbauer	Urban Operations Supervisor	Planning	35 years old	Higher Education Administration
Nelson Rogério ***Stahelin***	Administrative and Financial Manager	Administration and Finance	25 years	Higher Education Administration
Jacson Nienkotter	Planning Supervisor	Planning	10 years	Higher Education Administration
Valdir Juvenal Brazil Vasco	Analyst Logistics	Planning	15 years	Higher Level Industrial Design

Source: Author (2017).

Both the positions and the sector in which these professionals work lend credibility to the opinions obtained in the interview, as they reveal that they work in the area of the research (public transport operations). However, what adds most value to their opinion is the length of time they have worked in the area. These periods, provided by the interviewees, refer to the years they have worked in Joinville's public transport operating companies plus the years they have worked with Passebus. About three years ago, the professionals from the planning sector were brought together to work on the same system. The current administrative and financial manager has been with Passebus for 11 years, but before that he also worked for one of the operators.

6.3. RESULTS OF THE INTERVIEW

When asked about the participation of Passebus' Operational Planning Sector in projects such as those in the proposal to build bus corridors in Joinville, the professionals said that they are usually consulted when there is a major change that will affect bus routes. Generally, employees from the sectors take part in meetings in which the public planning body, currently SEPUD, presents the possible changes and asks Passebus for data such as journey times, number of journeys, how many passengers pass through the area, among others.

Asked specifically about the changes that would occur with the implementation of the exclusive corridors on the Beira-rio and Procópio Gomes axes, the professionals explained that they were only aware of the northbound corridor located on Avenida Beira-rio. The same northbound corridor has been in operation since the end of 2015, and with the proposal in place this axis would be used in both directions, north-south. The professionals said that there was a last meeting with SEPUD about 12 months ago, but until the date of the interview (16/10/2017), they had no knowledge of the Procópio Gomes axis. They also said that they had "heard" about a possible dual carriageway between Rua Procópio Gomes and Rua Urussanga, but this was about four years ago.

The professionals pointed out that, as seen in the analyses in section 5.3, the implementation of a corridor on Rua Procópio Gomes would result in speed gains for the operation, as this axis is currently a bottleneck in the system. On the Beira-rio axis, the exclusive northbound corridor solved an old problem. The use of this corridor has **provided "a great gain in quality on** the **lines", especially** on the Tupy-North line, one of the system's most demanding lines.

The professionals also pointed out that, when analysing the current number of preferential lanes in the city, the majority are in the north of the city. In the south, there are practically no lanes, with the exception of São Paulo and Santa Catarina streets and Getúlio Vargas Avenue. When you look at Avenida Coronel Procópio Gomes, Rua Florianópolis and Rua Monsenhor Gercino, where lines that serve the Paranaguamirim region pass through, with a high concentration of users, you can see that there is no corridor. These points made by the professionals not only highlight the importance of the Procópio Gomes axis, but also show the importance of the continuity of the corridors, since according to the proposal studied, the exclusive corridor ends at the beginning of Rua Florianópolis and does not serve Rua Monsenhor Gercino.

Given the importance of these corridors, they asked what changes would be made if they were implemented. The professionals pointed out that the first criterion to be analysed is the increase in operational speed, as this is what will bring the most results to the system. The second would be service, because if a line was changed, it would be necessary to check what the possible impacts would be in the region. They also pointed out **that depending on the region "it's not a good deal" to withdraw the line if it has already been serving that demand for a long time.**

They asked about the use of the company's database to monitor the growth in demand on certain lines. The professionals explained that this data is obtained from the ticketing system and the need for interference on one line or another is checked. They also pointed out that the database makes it possible to determine peak times, for example.

The professionals pointed out that currently, when satisfaction surveys are carried out, one of the main items that users consider relevant is the speed of the system, which has been gradually reducing. According to them, the corridors that are **implemented "are born to make up for lost time", in other words, they only make up for** lost **time,** and there is no evolution. As an example, they mentioned the recently installed corridor on Rua Nove de Março, which took so long to be implemented that the hoped-for gains didn't have much effect.

They pointed out that with the implementation of the corridor on Rua Nove de Março, other traffic lights appeared in the city, as well as problems with **intersections,** showing the treatment given to **intersections. And with that, they emphasised that a "real corridor" would suffer** less interference from other cars.

Once again, with regard to the sector's participation in corridor planning, the professionals reaffirm that the changes are presented to them and that they are involved, but that the project has already been drawn up when it is presented to them. From then on, when the project is implemented, the need for adaptation begins to emerge, such as changing the timetable, changing the line, in short, the implications of the changes are studied. Any adaptation or change to the operation suggested by the operational planning sector requires authorisation from SEINFRA (the Urban Infrastructure Department).

Some of the adaptations suggested by the sector are schedule reductions or line restructuring due to low demand. The data analysed for these changes is obtained from ticketing, as mentioned above, but the sector also receives suggestions from users via the SAC (user service), bus drivers and operation technicians who are present in the terminals on a daily basis.

The reduction in demand, also dealt with in the case study, was addressed in the interview. The professionals **said that the system "has a very large reduction in passengers every year". They pointed out that last year there was a drop of** around 8 per cent and this year, up to the date of the interview, there has been a drop of almost 6 per cent, which is worse than last year. For this reason, the operational planning sector must constantly seek to restructure the system and optimise the lines. The professionals say that sometimes it is necessary to merge two lines with low demand in order to reduce the cost of the system, because with the drop in users the cost of the operation increases.

They then asked about the frequency of the lines and the decisions made in relation to them. The professionals explained that there are several overlapping lines, different **lines on the same axis, due to the design of the city where the "main" streets have** several branches, similar to a herringbone, characterising the trunk-feeder system (reaffirming descriptions of the system made in the case study). They point out that the roads through which the trunk lines pass have more shops, offices, among other traffic generators, and end up requiring a higher frequency of service.

Given the structure of the system, they asked about timed integrations and users switching from a feeder line to a trunk line, with the aim of encouraging integrations for greater fluidity by making it possible for only trunk lines to use the main axes. In this respect, the professionals pointed out that there are currently few cases in which **users are forced to integrate, because, according to them, users don't have the "culture" of** using the ideal card (which enables timed integrations). If the percentage of users using the ideal card was close to 100 per cent, integration could be more widely used and there would be a wider range of possibilities for working with the lines.

Due to the importance of user information to facilitate integration over time, as seen in the case study, they asked about the realisation of a line map for the city, with the aim of identifying bus stops and integration possibilities. They said that they had already talked about it, but nothing concrete had happened. They also pointed out that changes to lines occur very frequently and would hinder the development of maps. For the time being, information on lines can be found on mobile phone apps, Google Maps and the Onibus.info website.

When asked about the information at the bus stops, they talked about the experience they had in Rua São Paulo, where metal and glass bus stops were installed, which provided panels with timetables and line identification, as well as advertisements as part of the subsidy to maintain the bus stop, but they said that they had all been vandalised. When asked how much they earn from advertising to maintain the bus stops, they said it's not enough. As a result, investment in the bus stops becomes unviable or insufficient due to maintenance costs.

The professionals said that they are currently studying with the City Council how to update the ticketing system, install GPS in the vehicles to obtain information in real time, among other possibilities. However, bringing in some of the technological and innovative tools currently available is a distant reality for Joinville. To do this, it is necessary to analyse certain difficulties, infrastructure issues and demand, thinking about the return on investment.

For them, what will really bring users back is progress in terms of speeding up the system, quality tracks and cheaper fares. They believe that the fare will only decrease with a set of actions that improve the user's journey and discourage the use of individual transport.

During the interview, some of the points made in the analyses carried out throughout the study were reaffirmed by the professionals. At the start of the interview, the lack of iteration between public sector planners and system planners was highlighted, and this lack of involvement was also verified in section 5.3.2.1, item c, between professionals from different phases of the project. Another worrying fact is that operational analyses are only carried out when the project is implemented, as stated in the interview.

With the database generated by ticketing and by monitoring the day-to-day operation of the system, various studies could be carried out. The interview with the professionals showed that, as was seen in section 5.3.2, these studies have not been carried out. They said that they make data and information relevant to public transport planning available to the responsible body, but pointed out that they were not aware of the specific changes that come from this data and information, or how it is used. The data and information they cited can be used for both line dimensioning and capacity calculations, but neither of these was carried out in the proposal analysed.

Both the professionals' view and the comparison with the MBRT document emphasised the importance of demand and speed for the system's efficiency. The interview pointed out the demands of the region to be served by the project and the importance of gaining operational speed for it. In terms of efficiency, the opinion of the professionals also emphasised the need to identify bottlenecks and deal with traffic lights, a topic discussed in section 5.3.2.3.

As pointed out in the same section, exclusive lanes are generally implemented where traffic congestion is a problem, or where congestion is likely to occur in the near future. Otherwise there would be no benefit in building a segregated bus lane. This is a condition pointed out by the professionals when they identified the Eixo Procópio Gomes as the system's current bottleneck.

With regard to time integration, line design and information for users, the point made in section 5.3.2.1, item d, is emphasised, as there is no map of the integrated system for users and the professionals themselves point out that it is necessary to use mobile applications to find information on line identification and timetables. They also pointed out the low take-up of the ideal card, which is necessary for time integration, showing that the lack of information not only makes it difficult to make transfers outside the terminal, but also discourages them. This behaviour makes the journey costly for the user, both financially and in terms of travel time, because in a trunk feeder system, which is the situation in Joinville, transfers are inevitable.

User dissatisfaction was identified early on in the study when the population's demands were pointed out. The professionals recognise the users' desire for a more efficient system and confirm that their use of public transport is falling. Finally, the professionals say that the fare will only decrease with a set of actions that improve the user's journey and discourage the use of individual transport, highlighting, as they did throughout the case study, the importance of the proposal to build exclusive bus lanes.

CHAPTER 7

FINAL CONSIDERATIONS

Disorganised growth in medium-sized cities has led to a demand for improvements in urban public transport, especially when you look at the drop in public transport users and the migration to individual transport. In recent years, SIT users in Joinville, classified as a medium-sized city, have been dissatisfied with the price of fares and the service provided. Given the decline in the number of Joinville public transport users and the population's demand for a higher quality and more efficient system, existing projects were analysed with a view to a fast, comfortable and cost-efficient transport system.

The main aim of this study was therefore to analyse the proposal for the construction of bus corridors made by Joinville City Council and submitted to PAC 2 MMC, comparing it to the PQBRT and MBRT documents. This was made possible by developing a case study in which the object of study was characterised, the proposal was elucidated and analysed in the light of the guiding documents for a BRT system.

As for the specific objectives, the first was to study the proposal for exclusive corridors presented by Joinville City Hall to PAC 2 MMC, and this was done in particular by explaining the programme, describing it, highlighting the axes selected for study and highlighting the demands of the population of Joinville. The second specific objective was to check whether the system fits the classification of a BRT system according to the criteria established in the PQBRT document, which was achieved using the scoring guidelines set out in the document, which assess different categories within a BRT system.

Analysing the proposal based on the PQBRT document showed that although the project does not have the basic elements to be characterised as a BRT system, the final score, obtained by applying the scoring guidelines, showed a result close to the minimum for a BRT system to be considered Bronze. This shows the potential of the corridors studied.

The third specific objective was to analyse whether the proposal complies with the guidelines presented in the MBRT document and this was done by identifying the type of access to the corridors, the service provided, the lines and the transfers. From these analyses it was possible to identify that the act of limiting access to operators, unlike what currently happens, will enable greater fluidity in the system, as there will be less interference from mixed traffic, thus leading to an increase in average speed and a reduction in users' journey times. It was also pointed out that the service that will be provided on the bus corridors, due to the operational logic of the trunk-feeder system, will meet three conditions that the MBRT document sets out for its success: high demand, differences in population density between different areas of the city and long distances between the city centre and the feeder areas. It should be noted that the axes studied have the potential for the service to be efficient, but that access to the terminals will be via preferential (non-exclusive) corridors, reducing the efficiency of the service.

The fourth and final specific objective was to verify the perception of urban public transport professionals in relation to the study's analysis. This was done by interviewing professionals in the field. In this interview, we obtained the professionals' perspectives on the researcher's perception and on what was observed during the analysis.

In summary, the analyses carried out throughout the work, both in the light of the MBRT and PQBRT guiding documents and by consulting professionals in the field, indicate that the proposal to build bus corridors has the potential to be efficient and bring improvements to Joinville's SIT. However, they also highlight the lack of studies such as the design of lines and networks, the identification of bottlenecks and traffic light control. The lack of these, coupled with the construction of preferential corridors linking the terminals, will probably jeopardise the quality of operation of the exclusive bus corridors proposed by Joinville City Hall.

7.1. LIMITATIONS ENCOUNTERED

Due to the lack of studies guided by the MBRT document, as well as the focus only on infrastructure in the existing studies in the proposal, it was necessary to search SEPUD's internal system to find possible studies dating from the year the proposal was made. The non-standardised nomenclature of the files and the lack of division by year in some folders made it difficult to find the documents.

The constant changes to the projects, even after the proposal was submitted to the programme, made it difficult to define the object of study. It was decided to analyse the original proposal, which is in the Secretariat's system, even though we know that there will be changes to its profile, as the original projects include improvements to the bus stops and their surroundings, and ensure the segregation of the corridors. However, given the date of the proposal in 2015, it is known that the costs of the work will change and to meet the proposed budget there will be simplifications in the projects, which could compromise the efficiency of the corridor once it is implemented.

7.2. SUGGESTIONS FOR FUTURE WORK

For future work, a study is suggested after the implementation of the analysed proposal, in order to identify the operational efficiency of the bus corridors, compare the score resulting from this work according to the guidelines set out in the PQBRT document and take into account the operating deductions provided by the same document.

Another relevant analysis would be of the existing bus corridors in Joinville, with the possibility of a diagnosis of their operation in the light of the guiding documents used in this work.

Finally, a suggestion is to analyse an existing BRT system in Brazil that is considered successful, in order to identify whether it complies with the guidelines set out in the guiding documents studied.

CHAPTER 8

REFERENCES

AFFONSO NETO, EMILIANO. **Urban mobility**. São Paulo: Secovi University, 2013. 44 slides, colour.

ANDRADE, Antonio Maria Claret Reis de. **Urban passenger transport modes.** 2015. Available at: <https://ie.org.br/site/ieadm/arquivos/arqnot9581.pdf>. Accessed on: 19 September 2017.

ANTP. **Panorama of urban mobility in Brazil:** Trends and challenges. São Paulo: Antp, 2006. 120 f

BRANCO, Soraia Patrícia Videira Martins. **Study and application of BRT systems - Bus Rapid Transit.** 2012. 172 f Dissertation (Master's) - Civil Engineering Course, Faculty of Engineering, University of Porto, Porto, 2013.

BRT BRASIL. **BRT Brasil Programme.** 2016. Available at: <http://brtbrasil.org.br/index.php/brt-brasil#.WhmxIVWnHIW>. Accessed on: 23 September 2017.

CARVALHO, Carlos Henrique Ribeiro; PEREIRA, Rafael Henrique Moraes. **Effects of fare variation and population income on the demand for urban public transport in Brazil**. Brasília: Institute for Applied Economic Research (IPEA), 2011. 39 f.

CASTRO, Cláudio de Moura. **The practice of research.** São Paulo: Mcgraw-hill do Brasil, 1977.

CASTRO, Felipe. **BRT or tram: a matter of choice?** Mobilize Brasil republishes text to clarify the main characteristics and the qualities and advantages of each mode of public transport. 2013. Available at: <http://www.mobilize.org.br/noticias/1331/brt-ou- vlt-questao-de-escol escolha.html>. Accessed on: 21 September 2017.

CNT (Brasília). **Urban buses face debt and falling demand.** 2017. Available at: <http://www.cnt.org.br/Imprensa/Noticia/onibus-urbanos-endividamento- queda-demanda>. Accessed on: 26 Nov. 2017.

COSTA, Priscila Bahia et al. Evaluation of the public transport system using the urban mobility index. **Revista dos Transportes Públicos: ANTP,** São Paulo, v. 145, n. 39, p.49-62, 30 May 2017.

CUCCI NETO, João. **Traffic lights: to be intelligent or not to be intelligent?** A comparison between fixed-time and real-time traffic light control. 2015. Available at: <http://meusite.mackenzie.br/professor_cucci/texto2.pdf>. Accessed on: 1 Nov. 2017.

EMBARQ BRASIL. **QualiÔnibus satisfaction survey**. 2014. EMBARQ Brasil, Joinville.

FERRAZ, A. C. P.; TORREZ, I.G.E. **Transporte público urbano.** 2. ed. São Carlos: Rima, 2004. 410 f.

IPPUJ - Fundação Instituto de Pesquisa e Planejamento para o Desenvolvimento Sustentável de Joinville (Org.). **Demographic density 2010**. Joinville: City Hall, 2010a.

IPPUJ - Fundação Instituto de Pesquisa e Planejamento para o Desenvolvimento Sustentável de Joinville (Org.). **CTA Joinville Map**. Joinville: City Hall, 2010b.

IPPUJ - Fundação Instituto de Pesquisa e Planejamento para o Desenvolvimento Sustentável de Joinville (Org.). **Economic and socio-environmental justification for interventions on the eastern axis - proposal 01 - PAC medium mobility cities**. Joinville: City Hall, 2013.

IPPUJ - Fundação Instituto de Pesquisa e Planejamento para o Desenvolvimento Sustentável de Joinville (Org.). **Joinville's mobility and accessibility plan (PlanMOB).** Joinville: City Hall, 2014a, 107 f.

IPPUJ - Fundação Instituto de Pesquisa e Planejamento para o Desenvolvimento Sustentável de Joinville (Org.). **Map of the PAC 2 DEZ 2014 programme with anemones**. Joinville: City Hall, 2014b.

IPPUJ - Fundação Instituto de Pesquisa e Planejamento para o Desenvolvimento Sustentável de Joinville (Org.). **EIXO LESTE**. Joinville: City Hall, 2014c.

IPPUJ - Fundação Instituto de Pesquisa e Planejamento para o Desenvolvimento Sustentável de Joinville (Org.) **Figure 01 - Proposed repositioning of boarding and disembarking points**. Joinville: City Hall, 2014d.

IPPUJ - Fundação Instituto de Pesquisa e Planejamento para o Desenvolvimento Sustentável de Joinville (Org.). **Preliminary Notebook: Joinville Sustainable Mobility Plan (PlanMOB)**. Joinville: City Hall, 2015a, 119 f.

IPPUJ - Fundação Instituto de Pesquisa e Planejamento para o Desenvolvimento Sustentável de Joinville (Org.). **José Vieira executive project**. Joinville: City Hall, 2015b.

IPPUJ - Fundação Instituto de Pesquisa e Planejamento para o Desenvolvimento Sustentável de Joinville (Org.). **Procópio Gomes executive project**. Joinville: City Hall, 2015c.

ITDP BRAZIL. **BRT Quality Standard**. 3 ed. 2016. Institute for Transport and Development Policies, Rio de Janeiro.

KREIDLOW, Rogério. **Joinville's Beira-river avenue to be changed to a motorway BRT:** Work on the Bus Rapid Transit, which begins next year, is part of a R$104.8 million investment by PAC 2 and will upgrade 55km of roads. 2013. Available at: <http://anoticia.clicrbs.com.br/sc/geral/noticia/2013/10/avenida-beira-rio-de-joinville- muda-para-ter-corredor-de-brt-4292733.html>. Accessed on: 05 Sep. 2017.

LOBO, Renato. **Trolleybus or Hybrid Bus? See how the modes compare.** 2014. Available at: <http://viatrolebus.com.br/2014/08/trolebus-ou-onibus-hibrido-veja-a- comparacao-dos-modais/>. Accessed on: 21 September 2017.

MINISTRY OF CITIES. **BRT Manual - Planning Guide**. 3. ed. 2008. Ministry of Cities, Brasilia.

MINISTRY OF CITIES. **National Sustainable Mobility Policy**. Brasilia: Ministry of Cities, 2004.

MINISTRY OF CITIES. **Medium-sized Cities Mobility PAC to be announced.** 2012. Available at: <http://www.pac.gov.br/noticia/f7ae8e88>. Accessed on: 05 September 2017.

NTU. **NTU Yearbook:** 2016-2017. Brasília: National Association of Urban Transport Companies, 2017, 70 f.

ONIBUS.INFO (Joinville). **BUS LINES:** Urban public transport lines in Joinville, Santa Catarina. Available at: <https://onibus.info/linhas>. Accessed on: 10 October 2017.

PASSEBUS (Joinville). **Institutional.** Available at: <http://www.passebus.com.br/>. Accessed on: 5 Nov. 2017.

POLIAK, Milos et al. The Competitiveness of Public Transport. **Journal Of Competitiveness,** [s.l.], v. 9, n. 3, p.81-97, 30 Sep. 2017. Tomas Bata University in Zlin. http://dx.doi.org/10.7441/joc.2017.03.06.

JOINVILLE CITY HALL. **City Hall signs R$105 million contract for mobility investments:** Financing with Caixa Econômica Federal will allow Joinville's largest road investment. 2015. Available at: <https://wwwold.joinville.sc.gov.br/noticia/9111-City Hall+signs+contract+of+R$+105+million+for+investments+in+mobility.html> . Accessed on: 05 Sep. 2017.

PRESTON, J. M. Public Transport. In: KITCHIN, Rob; THRIFT, Nigel (Ed.). **International**

Encyclopedia of Human Geography. Elsevier Science, 2009. p. 452-459. Available at: <https://eprints.soton.ac.uk/75839/>. Accessed on: 12 October 2017.

RECK, Garrone. **Workbook:** Public transport. Curitiba: Federal University of Paraná, 2017. 108 p. Available at: <http://www.dtt.ufpr.br/Transporte Publico/Arquivos/TT057_Apostila.pdf>. Accessed on: 08 September 2017.

RODRIGUES, Janda Cecília Ribeiro; SOUZA FILHO, João Pedro de; GOMES, Nayara Paula. **Preferential Corridors - BRS as a mechanism for gaining operational speed in public transport:** a case study for Goiânia. 2014. 72 f. TCC (Graduation) - Civil Engineering Course, Universidade Federal de Goiás, Goiânia, 2014.

SEPUD - Joinville Urban Planning and Sustainable Development Department. Joinville City in Data 2017. Joinville: City Hall, 2017a, 73f.

SEPUD - Joinville's Department of Urban Planning and Sustainable Development. **Invitation to Tender No.##/ of 01 September 2013:** Annex II - Basic Project. Document available for consultation only. Joinville, City Hall, 2017b.

SILVEIRA, Márcio Rogério; COCCO, Rodrigo Giraldi. Public transport, mobility and urban planning: essential contradictions. **Estudos Avançados,** São Paulo, v. 27, n. 79, p.41-53, oct. 2013. Available at: <https://www.revistas.usp.br/eav/article/view/68701/71281>. Accessed on: 12 October 2017.

SOUSA, Marcos de. **BRT, tram or monorail: which is the best medium-capacity solution?** In times of renewed investment in public transport, we've put together some texts comparing the three medium-capacity transport systems. 2014.
Available at: <http://www.mobilize.org.br/noticias/7161/brt-e-vlt-qual-e-a-melhor- solucao.html>. Accessed on: 21 September 2017.

VASCONCELLOS, Eduardo Alcântara de; CARVALHO, Carlos Henrique Ribeiro de; PEREIRA, Rafael Henrique Moraes. **Transport and urban mobility**. Brasília, DF: ECLAC. Brazil Office/ IPEA, 2011 (Textos para Discussão CEPAL-IPEA, 34), 74f.

WRI BRASIL SUSTAINABLE CITIES (Porto Alegre). **Joinville (SC).** Available at: <http://wricidades.org/nosso-trabalho/projeto-cidade/joinville-sc>. Accessed on: 15 October 2017.

CHAPTER 9

ANNEXES

Annex 1 - Map of the PAC 2 DEC 2014 programme with anemones.

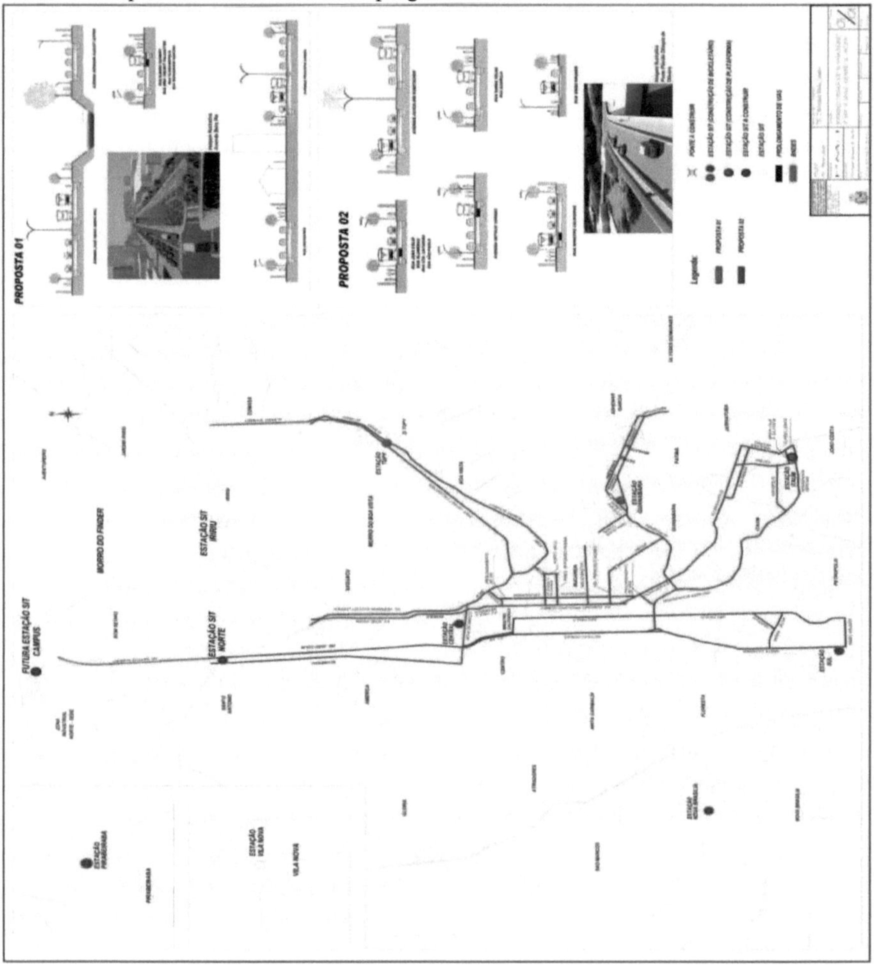

Annex 2 - Proposed repositioning of boarding and alighting points.

FIGURA 01 - PROPOSTA DE REPOSICIONAMENTO DOS PONTOS DE EMBARQUE E DESEMBARQUE

Annex 3 - CTA Joinville map.

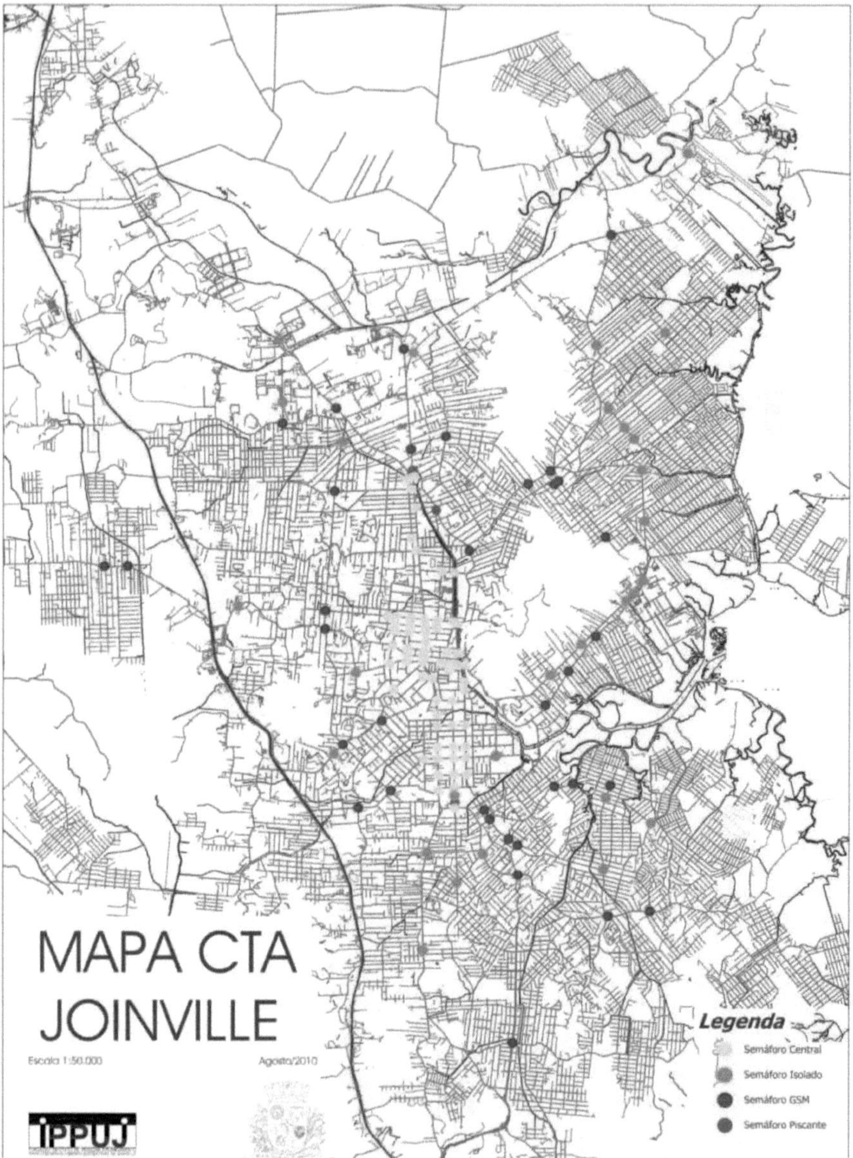

APPENDIX

Interview script

Company visited: Passebus.
Name of interviewee:
Position of interviewee:

1. What study is carried out for decisions on creating new lines or changing routes?
2. How do you determine the ideal vehicle and the necessary frequency for a line?
3. How does time integration work at bus stops?
4. Were there any network and line projects for the implementation of the exclusive corridors proposed for PAC 2 MMC? If so, which organisations were involved? Is it possible to access this document?
5. What future projects do you have to improve user access to information? Have you considered designing the lines to identify the numbering, direction and the possibility of transferring? What is the company's view of the need for information in the shelters?

MIX
Papier aus verantwortungsvollen Quellen
Paper from responsible sources
FSC® C105338

Printed by Books on Demand GmbH, Norderstedt / Germany